FIELD MANUAL
NO. 21-18

HEADQUARTERS
DEPARTMENT OF THE ARMY
Washington, DC, 1 June 1990

FOOT MARCHES

DISTRIBUTION RESTRICTION: Approved for public release; distribution is unlimited.

★ This publication supersedes FM 21-18, 29 January 1971.

CONTENTS

Page

PREFACE

This manual is a guide for commanders and their staffs in the procedures and techniques of foot marches. It describes the march mission, characteristics and types of foot marches, and march training to include planning procedures, duties of commanders, march discipline, march hygiene, and march safety. The material herein applies to all levels of conflict without modification.

This manual is for use with other field manuals where modification of foot marching procedures and techniques for specific regions of the world is required. Details concerning operations in desert, jungle, northern, and mountain areas are contained in FM 90-3, FM 90-5, FM 31-71, and FM 90-6, respectively. Tactical foot marches are discussed in FMs 7-10 and 7-20.

The provisions of this publication are the subject of the following international agreement:

STANAG 2154 Regulations for Military Motor Vehicle Movement by Road

When amendment, revision, or cancellation of this manual is proposed, which will effect or violate the international agreement concerned, the preparing activity will take appropriate reconciliation action through international standardization channels to include the departmental standardization officer.

The term "company" as used herein can also mean battery or troop.

The proponent of this publication is HQ TRADOC. Submit changes for improving this publication on DA Form 2028 (Recommended Changes to Publications and Blank Forms), and forward it to Commandant, US Army Infantry School, ATTN: ATSH-ATD, Fort Benning, GA 31905-5410.

Unless otherwise stated, whenever the masculine gender is used, both men and women are included.

CHAPTER 1

Introduction to FOOT MARCHES

Foot marches are the movement of troops and equipment mainly by foot with limited support by vehicles. They are characterized by combat readiness, ease of control, adaptability to terrain, slow rate of movement, and increased personnel fatigue. Foot marches do not depend on the existence of roads.

1-1. HISTORICAL EXAMPLES

Many examples of successful marches exist throughout the history of warfare.

a. A good example of a successful march occurred during World War II. It was the grueling foot march during the Sicilian campaign from 20 to 21 July 1943. The 3d Battalion, 30th Infantry Regiment, 3d Infantry Division performed this march. The battalion was directed to move on foot across mountains from Aragona to San Stefano to enter into a coordinated attack on enemy forces in San Stefano. The battalion made this record-breaking, 54-mile, cross-country march in only 33 hours due to continuous marching. Two hours after arrival, the battalion was committed in the attack on San Stefano, which resulted in its capture.

b. A second example was the movement of large elements of the 3d US Army during the battle of Ardennes in 1944 to stop.

the enemy counteroffensive. On 16 December 1944, while the 3d US Army was preparing to attack the Siegfried line in Germany, the Battle of the Bulge commenced. By 19 December, the German attack had reached such large proportions that the 3d US Army was directed to cease its attack to the east and to turn north. The 3d US Army shifted its troops from the Saarlautern-Saarbrucken area to the Luxemburg-Belgium area, a distance of 100-road miles. The III Corps launched the new attack at 0600 on 22 December 1944.

(1) The size of this operation is indicated by the statistics. Eight divisions were moved from the eastern to the northern sectors, and two divisions were brought in from the rear areas. This shifted about 250,000 men and 25,000 vehicles. Units moved distances varying from 50 to 150 miles. From 18 to 31 December 1944, the trucks of the 3d US Army traveled a total of 1,254,042 miles. From 17 to 23 December 1944, nearly 42,000 tons of supplies were moved north. In the same week, seven traffic control stations in the area checked through 133,178 vehicles. The weather was damp and cold, and the ground was covered with snow and ice. However, foot soldiers marched to entrucking and from the detrucking locations to their new battle areas.

(2) This decisive movement of an entire army was possible since smaller units could move themselves. The units of the 3d US Army were well trained in movement. Their SOP and experience tables had been tested and proved. When the job was placed upon them, they succeeded, knowing the importance of combat readiness.

1-2. FUNDAMENTALS

Troop movement is the transporting of troops from one place to another by any available means. This is inherent in all military operations. A successful move places troops and equipment at their destination at the proper time ready for combat. Troop movement is conducted by foot or motor marches, by rail, by air, or by water, or by various combinations of these methods.

a. Detailed plans are required for effective troop movement. However, units must be notified early of an impending move to allow for effective planning.

b. A successful troop movement also depends on effective control of units during movement. Such control is accomplished through the chain of command, and by proper supervision and organization of the force.

1-3. MARCH MISSION

A successful foot march is when troops arrive at their destination at the prescribed time. They are also physically able to execute their tactical mission.

a. Troops must execute the mission immediately upon completing the march. Normally, this is done through conditioning and acclimatization of troops to the area of operations. This includes physiological and psychological adjustment by the individual soldier.

b. Commanders must ensure that the amount and type of equipment carried, the rate of march, and the length and number of rests equates with the physical endurance of the men. Good planning and command leadership are required to move troops to the right place at the right time. The commander also ensures troops arrive in good condition to accomplish their mission.

1-4. CLASSIFICATION OF MOVEMENTS

Troop movements can be tactical or administrative. Both classifications apply in most movements but one is normally dominant.

a. **Tactical.** Tactical movements are conducted in the combat zone to emphasize tactical considerations such as security and the use of combat-ready formations. They reemphasize efficiency and ease of movement, and they anticipate ground contact with the enemy en route or after arriving at the destination. Movements may be conducted over unsecure routes if there are no friendly forces between the foremost elements of the moving force and the enemy (Figure 1-1).

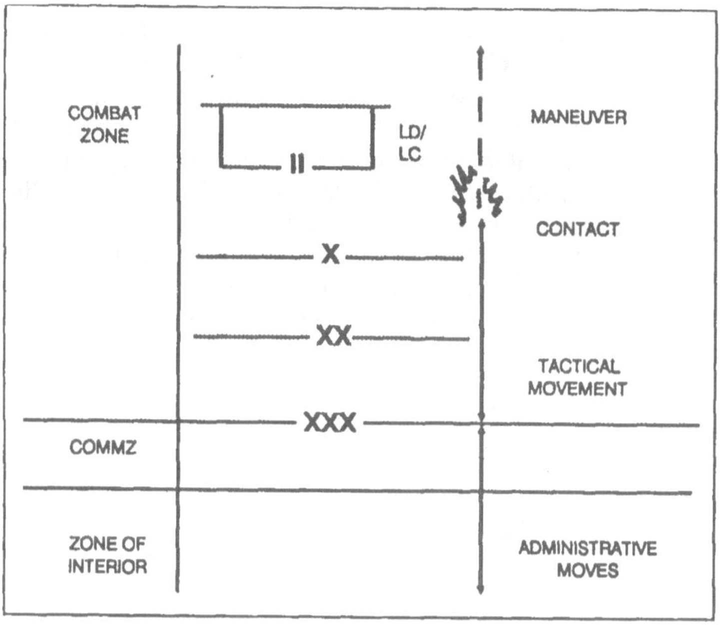

Figure 1-1. Movement.

(1) When relocating in the combat zone, the unit conducts tactical foot and motor marches in the division and corps rears. This occurs before hostilities begin or when a forward defense has been established. Speed is vital, and security requirements are minimal.

(2) Units move by tactical foot or motor marches to an assembly area where they prepare to conduct combat operations. During tactical movements, the commander must be prepared to maneuver against an enemy force.

(3) Once a unit is deployed in its assigned zone or sector, it normally moves using the proper techniques for the assigned mission. When contact is made, tactical movement becomes maneuver.

b. **Administrative.** Administrative movements are conducted in the COMMZ and zone of the interior. They emphasize the best method of movement and reemphasize tactical considerations. Administrative movements are based on likely ground contact with the enemy being remote, both en route and soon after arrival at the destination. They are normally conducted over secure routes if friendly forces are between the foremost elements of the moving force and the enemy. When conducting administrative movements, units should maintain integrity and practice security techniques that pertain to tactical road marches.

1-5. TYPES OF MARCHES

Each type of march has its own purpose and application. Those discussed herein are common to both administrative and tactical marches.

a. **Day Marches.** In the absence of enemy threats, day marches are preferred. They permit faster movement and are less tiring for troops. They are characterized by dispersed formations, ease of control and reconnaissance, and increased vulnerability to enemy observation and air attack.

b. **Limited Visibility Marches.** Limited visibility marches are characterized by closed formations, difficult command and control and reconnaissance, slow rate of march, and good concealment from hostile observation and air attack. Night marches also exploit darkness to gain surprise and help units avoid extreme heat common to day marches. Control of the march requires detailed planning, stringent control measures, thorough training, and march, light, and communication disciplines.

(1) If concealment is required, movement before dark is restricted to small detachments. Also, the march should be completed by daybreak with troops in concealed positions. When movement is near the enemy, noise suppression must be considered. To conceal operations from the enemy and to

prevent him from gaining information about the march, measures to maintain security must be enforced. The use of navigational aids, such as ground surveillance radar, thermal sights, and night vision devices, can prevent many command and control problems encountered during night movements.

(2) Limited visibility marches must be carefully planned. This includes reconnoitering of routes and assembly areas. Special precautions ensure direction and contact within the column. Therefore, guides and file formations are needed.

c. **Forced Marches.** Forced marches require speed, exertion, and more hours marched. They are normally accomplished by increasing the marching hours for each day rather than the rate of march. Forced marches are employed only when needed since they decrease the effect of units. Troops should be told the reason for a forced march to ensure maximum effort.

d. **Shuttle Marches.** Shuttle marches alternate riding and marching in a troop movement. This is normally due to lack of enough vehicles to carry the entire unit. Shuttling requires transporting troops, equipment, and supplies by a series of round-trips with the same vehicles. It can be performed by hauling a load an entire distance and then returning for another. It may also be performed by carrying successive parts of a load for short distances while the remaining parts continue on foot.

CHAPTER 2

Factors Affecting FOOT MARCHES

Factors that greatly influence the conduct of the march are the location of enemy forces, the nature of the terrain and weather, and the activity of enemy aviation. Contact with the enemy should be expected from any direction that is not protected by friendly forces or terrain barriers. Other factors that commonly affect the conduct of a march include the distance to be marched; planning effectiveness; march discipline and supervision; time available; and physical condition, training status, and attitude of the soldier.

2-1. TACTICAL CONSIDERATIONS

Soldiers should not be required to walk until they enter battle. From that point on, they should be required to carry only what they wear, weapons, ammunition, water, and minimal rations.

a. A moving force protects itself against enemy action by the use of all-round security elements, which must be employed during all types of movements. These security forces provide early warning of enemy threats, and are organized and arranged IAW the mission and enemy capabilities.

(1) As ground contact with the enemy becomes imminent, large units are normally preceded by a covering force, operating directly under the senior commander. The covering force is a self-contained tactical unit that develops the situation, seizes key

terrain features, and defeats enemy resistance. Typically, divisions and larger units establish a covering force.

(2) Each column commander, even if preceded by a covering force, also establishes an advance guard with the mission of protecting the main body from surprise, ensuring its uninterrupted advance. Flank and rearguards, operating directly under the column commander, protect unsecured flanks and rear.

b. Before beginning the march, routes should be reconnoitered and marked, bridge capacities indicated, and dimensions of underpasses compared to vehicle dimensions (if the tactical situation permits). Timely measures are enforced for the reduction of obstacles and other possible causes of delay.

c. Combat support assets are located in the column for quick availability to their units. Those not immediately required can be held to the rear and sent forward when the situation dictates. Sustainment loads not immediately needed for combat can be left in the old area and moved forward after the movement of troops, essential equipment, and supplies has been completed. Each grouping for movement corresponds with the organization for combat to include weapons, ammunition, and rations needed for combat at the destination.

d. Shuttle marches aid movement and conserve energy when there are insufficient vehicles to move the entire force at one time. Foot troops to be transported by vehicle in a later echelon can march to an intermediate point where they are met and picked up by vehicles returning from an earlier echelon, or foot troops transported in an earlier echelon can disembark short of their destination and march the rest of the distance while vehicles return for a later echelon. Plans for combined foot and motor movement must be flexible since circumstances in forward areas can easily disrupt the time schedule.

e. When marching to occupy an assembly area, the commander can consider conducting the march by infiltration; that is, marching in vehicles or on foot by small groups extended over time to give the impression of casual traffic. This technique is used to provide secrecy, deception, and dispersion in areas where enemy observation or attack is likely, but it requires more time, more detailed planning, and greater decentralization of control.

f. When the possibility of contact with enemy ground forces is remote, the principal objects are to facilitate and expedite movement, and to conserve the energy of troops. Columns consist of units having the same rate of movement. Those having different rates of movement are assigned separate routes, or their movements are staggered along the same routes.

g. When an unforeseen crossing of two columns occurs and no control personnel from a higher headquarters are present, the senior commander regulates the crossing, based on the situation and the missions of the two columns.

2-2. EFFECTS OF WEATHER AND TERRAIN

The varying types of terrain over which troops must march present different problems for commanders, depending on the specific area of operations. Weather conditions combined with terrain affect the mobility of marching troops.

a. Foot marching under adverse climatic conditions follows the same principles asunder normal conditions. The differences depend on the physical limitations imposed by adverse conditions and the use of special equipment to overcome them.

b. Restrictions imposed by weather extremes and terrain constitute the major change from operations intemperate areas. These restrictions can present major obstacles to the successful conduct of operations unless proper provisions are established.

c. Movement in many areas of the world must be calculated in terms of time and distance. The problems are how much time troops need to go from one place to the other and the distance between them. This applies mainly in arctic, mountain, or jungle areas where trails are either limited or nonexistent, and where cross-country movement can be slow and arduous.

2-3. MARCH DISCIPLINE

March discipline includes observing and enforcing the rules and instructions that govern a unit on a march, which include formation, distances between elements, speed, and the effective use of concealment and cover. It must also include specific controls and restrictions such as water, light, noise, and communication disciplines. March discipline is the culmination of effective training, which results in enthusiastic teamwork among all soldiers of the unit.

2-4. WATER DISCIPLINE

Water discipline must be observed by all unit members to ensure best health and marching efficiency. The following rules must be adhered to:

- Drink plenty of water before each march to aid sustainment during movement.
- Drink only treated water from approved sources.
- Drink water often. Water should be consumed before, during, and after the march.
- Drink small quantities of water rather than gulping or rapid intake.
- Drink water even when not thirsty.
- Drink water slowly to prevent cramps or nausea.
- Avoid spilling water.
- Refill canteens at every opportunity.

a. The human body does not operate efficiently without adequate liquid intake. When soldiers are engaged in strenuous activities, excessive amounts of water and electrolytes are lost through perspiration. More water is lost through normal body functions such as respiration and urination, which can create a liquid imbalance in the body. As a result, dehydration could occur unless this loss is immediately replaced and soldiers rest before continuing their activities. Deficient liquid and salt intake during hot weather can also result in heat injuries.

b. The danger of dehydration is as prevalent in cold regions as it is in hot, dry areas. The difference is that in hot weather the soldier is aware that his body loses liquids and salt through perspiration. In cold weather, when a soldier is bundled up in many layers of clothing, he has difficulty knowing this condition exists since perspiration is rapidly absorbed by heavy clothing or evaporated by the air—it is rarely visible on the skin.

c. Salt in food compensates for the daily salt requirement. Additional intake of salt should be under the direction and supervision of a physician or physician's assistant.

d. If pure water is not available, water in canteens can be treated by adding water purification tablets. *(See FM 21-10 for methods of purifying water.)*

e. If the unit is forced to traverse a contaminated area (NBC) due to the tactical situation, water consumption increases and forced hydration becomes necessary. Leaders at all levels must try to prevent heat injuries brought on by physical activity in an NBC environment.

2-5. ACCLIMATIZATION PROCEDURES

Soldiers must be physically and mentally conditioned to effectively participate in foot marches.

a. The many types of terrain and climate throughout the world require different acclimatization procedures for successful operations. Ideally, troops should be trained to operate in all areas with a minimum of preparation; however, each area has specific requirements that must be met before operating in it. For example, troops scheduled for operations in mountains would normally participate in high-altitude training for 10 to 14 days before engaging in full-scale mountain marches.

b. Psychological adjustment eliminates preconceived notions and fears about specific locations and climates. Training conducted logically and realistically causes most soldiers to lose previously held fears of height, cold, or isolation. The adjustment is facilitated by educational programs that gradually introduce soldiers to unfamiliar terrain features or climates. During this program,soldiers are encouraged to develop confidence until they can operate in these areas with ease and assurance.

c. Self-confidence in each soldier is a direct result of effective psychological adjustment. Therefore, soldiers can benefit from training since they believe in their abilities. Self-confidence in foot marching is developed by strong leadership and progressive training. As soldiers become stronger and as marching techniques are learned and applied, a soldier's self-confidence and pride increases. Leaders can also stimulate pride by building unit spirit and by instilling a determination to succeed. A well-planned and well-conducted march is an excellent way to develop and demonstrate the many attributes of a good soldier, a good leader, and a good unit.

NOTE: Specific adjustment factors and techniques are discussed in field manuals that relate to arctic, mountain, desert, and jungle operations.

2-6. MORALE

Morale can greatly affect the marching effectiveness of troops. Low morale can be contagious and magnifies any discomfort soldiers might experience. Leaders can improve morale by applying proper march and leadership techniques, some of which are discussed herein.

a. Provide advance warning of a march to the unit so troops can prepare adequately—more time should be allowed during early training.

b. Do not form the unit too early. Hold formation early enough to allow time for inspecting troops and for performing last-minute checks.

c. During the march, avoid delays that keep soldiers standing. Delays can increase fatigue and cause legs to stiffen, making it more difficult to resume the march. A route reconnaissance before the march provides information on conditions that could cause delays - advance action can prevent them.

d. Prescribe and maintain a steady rate of march—too rapid or too slow a rate induces fatigue.

e. During the march, hold passing vehicles to a reasonable speed to promote safety and to prevent dust, rocks, or mud from being thrown on soldiers. If dust conditions are offensive, move troops to the upwind side of the road.

f. Do not allow trucks to overtake the column that are used to transport stragglers or march casualties unless it is unavoidable.

g. Ensure soldiers to the rear of the formation receive a full breaktime.

h. Ensure leaders at all echelons march with their soldiers throughout the duration of the march. Soldiers quickly detect the presence or absence of their leaders in a foot march.

i. Encourage unit leadership to walk the length of the marching unit periodically to spotcheck soldier performance and well being, and to ensure command presence is observed.

j. Ensure availability of adequate water at rest stops and throughout the march.

2-7. INDIVIDUAL LOAD

To prevent an individual load from hindering a marching soldier's mobility and combat readiness, commanders must reduce the carried load to the minimum mission-essential and survival equipment.

a. The individual's combat load is that mission-essential equipment as determined by the commander, which is required for the soldier to fight and survive immediate combat operations. The load can be divided into an approach march load and fighting load. The fighting load should not exceed 48 pounds, and the approach march load (which includes the fighting load) should be less than 72 pounds, based upon individual abilities.

b. The primary consideration is not how much a soldier can carry, but how much he can carry without impaired combat effectiveness—mentally or physically. The combat strength of a unit cannot be counted solely by the number of soldiers but must be counted by the number of willing and physically able soldiers.

(1) Soldiers become exhausted quickly when under the stress of combat. Therefore, they must be required to carry less into battle than they are conditioned to carry during training. Soldiers should be conditioned for carrying heavy loads but should be equipped in combat to move swiftly.

(2) The individual load must not be based on the gear and supplies needed to meet every contingency. The commander should not expect his soldiers to carry enough gear for all possible combat situations. Instead, items to be contained in the load must be based on realistic expectations.

(3) The commander is responsible for obtaining the means to carry additional gear. Usually a rifle company or smaller-size unit requires one truck and one trailer in support to carry additional gear. In cold weather, or during other conditions where personal gear increases, the requirement increases.

(4) The commander must ensure that the supply system provides the balance of essential supplies and equipment that are not carried by the unit. *Soldiers must feel confident that their needs will be met.*

(5) In training, commanders must instill pride in their soldiers when operating under austere conditions. Soldiers should be trained in field craft techniques and the use of caches and field expedients. However, the commander must set the standards.

NOTE: See Chapter 5 for a detailed discussion on soldier's load management.

CHAPTER 3

Movement Planning for FOOT MARCHES

The success of the march depends upon thorough planning that must consider the mission, tactical situation, terrain, weather, and participating units. A successful march requires the unit to adhere to prescribed routes and time schedules, to effectively employ all available means of transportation, and to execute assign tasks immediately upon arrival at the destination. Movement planning culminates in the preparation and issuance of a standard operation order with required annexes—written movement orders are rarely prepared at company level.

This chapter implements STANAG 2154 (Edition Five).

3-1. PLANNING

March planning is based on that planning conducted at battalion level and may be organized IAW with the following sequence.

a. **Receive the Mission.** The unit receives the mission to conduct a road march. The planning process begins with the commander and staff conducting a hasty mission analysis to determine critical times and tasks to accomplish.

b. **Preparation and Issuance of the Warning Order.**To allow subordinate units the required time to prepare for a pending move, a warning order is issued, which contains all available information about the march. The planning time available determines the time of issuance and the content of the warning order.

c. **Estimate of the Situation.** In the operation estimate, the S3 considers the mission, weather, terrain, time and space factors, available routes, available transportation, enemy capabilities, disposition of own forces, physical condition and training of troops, and courses of action available. The S3 then recommends to the commander which course of action to accept. Based upon the commander's decision, the staff then prepares the movement order. *(See FM 101-5-1 for a detailed discussion of the estimate of the situation.)*

d. **Development of Detailed Movement Plans.**After the commander has selected a course of action, an OPORD is prepared. In developing these plans, the commander or staff must consider the following.

(1) *Column organization.* To facilitate control and scheduling, units are organized into serials and march units, and are given an order of march.

(a) A march unit is a unit of command that moves and halts at the command of a single commander. The march unit normally corresponds to one of the smaller troop units such as a squad, section, platoon, or company.

(b) A march serial (referred to as a serial) consists of one or more march units that are organized under the senior officer and are given a specific numerical or alphabetical designation to facilitate control. The march units of the serial normally possess the same march characteristics. A serial is usually a battalion or larger unit but can be a company if marching alone.

(c) A march column (referred to as a column) consists of elements of a command that are moving over the same route and can consist of one or more serial. A column commander is designated to facilitate control. A column is normally a brigade or larger unit but can be a battalion if marching alone.

(2) *Order of march.* In a tactical march, the order of march depends on the mission, terrain, probable order of commitment into action, and mobility of units. March units and serials are placed in the desired order of march by scheduling the arrival of march units at the start point.

(a) If tanks and infantry units are included in the march, they are interspersed throughout the column to facilitate integrated entry into combat.

(b) Artillery and mortars are placed forward and throughout the column to ensure the support of the security forces and the initial action of the main body.

(c) Air defense weapons are deployed throughout the column or are moved by bounds to protect passage of critical points.

(d) Engineer units are located well forward to facilitate the movement of the force through obstacles along the march route.

(e) Antitank weapons can be disposed to provide protection throughout the column. Some antitank weapons may be employed in support of security forces.

The integration of these and other combat, combat support, and combat service support assets may have an adverse impact on the movement of forces.

(3) *March formations.* The formation for foot marches varies depending on the routes available and the enemy situation. The usual formation for tactical marches is a column of two files with one file on each side of the road or in single file. The column commander designates the side of the road on which the troops are to march, or whether both sides of the road are to be used. Based on the enemy ground threat, the column forms into a route column, tactical column, or approach march.

(a) Route column. A route column is enforced when the likelihood of ground contact with the enemy is remote.

Administrative considerations govern movement; therefore, units are grouped administratively for ease of movement and control. Commanders normally move at the head of their units. This formation is sometimes called an *administrative column.*

(b) Tactical column. A tactical column is enforced when ground contact is possible. Units are grouped tactically to permit prompt adoption of combat formations. Movement is usually conducted over roads or trails and by the fastest means available. March units establish local security to the flanks. Dispersion depends on the enemy situation.

(c) Approach march. An approach march is enforced when ground contact with the enemy is imminent. Tactical considerations govern; therefore, elements whose contact with the enemy is likely adopt suitable combat formations. The commander's main concerns are to quickly bring superior combat power to bear against the enemy and to protect his force against surprise. The column establishes guards to the front, flanks, and rear, but larger forces should establish a covering force to ensure unimpeded movement.

(4) *March computations.* Based on the strength, formation, and rate of march, march unit pass time is computed. The pass time of the marching columns, plus necessary time-distance computations, is used to determine the completion time of the march.

(5) *Road movement table draft.* Using the march computations, a draft of the road movement table is completed.

(6) *Command and control.* The commander establishes initial control of the march by designating control measures in his road movement order. Examples of control measures are:

- Start point and release point.
- Other critical points along the march route (checkpoints, passage points, and so on).
- Time at which the head or tail of the column is to pass the SP and critical points.

- Rate of march.
- Order of march.
- Assembly or bivouac areas.
- Location of command post.
- Communications for use during the march.

The commander provides for advance and quartering parties, guides, route marking, and traffic control. Army aviation and military police units are particularly suitable for traffic control.

(7) *Plan check.* Using the draft road movement table and a road movement graph, the movement plan is checked to ensure that it conforms to the directive of the higher headquarters and the battalion commander's instructions.

(8) *Tactical situation.* The march order should also contain a statement of enemy situation, weather, and visibility conditions, and if applicable:

- Road restrictions.
- Information obtained from route reconnaissance.
- Actions on enemy contact (ground and air).
- Actions at halts and for disabled vehicles.
- Actions in the assembly area.
- Procedures for resupply, maintenance, and feeding.
- Location of leaders.
- A communications plan.

Most of the information should be part of the unit's SOP; therefore, only exceptions to the SOP should be stated in the OPORD.

e. **Issuance of Road Movement Orders.** The march order is prepared either as an OPORD or as an annex to an OPORD. (An example of a road movement order is contained in Appendix D.) The OPORD is either written or issued orally, and is accompanied by a road movement table, operation overlay, or strip map.

(1) A road movement table, prepared as an annex to an OPORD, provides serial commanders with arrival and clearance

times at checkpoints along the route of march. It provides the column commander with information as to the proposed location of elements of the column at various times.

(2) An operation overlay shows the location and strength of friendly forces involved in an operation and should show the present location of units, route of march, critical points, and the new location of units at the destination.

(3) A strip map is a schematic diagram of the route of march, and shows landmarks and checkpoints with the distances between them. It can be issued as an annex to the road movement order and in addition to or in lieu of an overlay.

(4) An administrative order or annex can be cited or included in the OPORD when the administrative details are too long for inclusion in the body of the order.

f. **Organization and Dispatch of a Reconnaissance Party.** Each march plan is based on a thorough ground reconnaissance, time permitting. Map reconnaissance and aerial reconnaissance help formulate a plan but are not substitutes for ground reconnaissance. A reconnaissance party performs the route reconnaissance and usually consists of a reconnaissance element, engineer element from an attached or supporting engineer unit, and traffic control element. When the situation dictates, NBC survey teams may be included in the reconnaissance party. The unit SOP establishes the composition of the reconnaissance party, which can be modified to meet the requirements of a specific march. The information required by the S3 from the reconnaissance party includes:

(1) Available routes and conditions (routes may be specified by higher headquarters).

(2) Recommended rate of march.

(3) SP and RP selections or confirmation of their suitability, which was selected by map reconnaissance.

(4) Confirmation of the assembly or bivouac area location.

(5) Checkpoint locations on the route.

(6) Distance between checkpoints on the route and total distance from SP to RP.

(7) Location of obstacles and estimation of soldiers and equipment needed to repair and maintain routes.

(8) Number of guides required and their route location.

g. **Organization and Dispatch of a Quartering Party.** Quartering party members prepare anew area for the systematic arrival of units.

(1) A battalion quartering party consists of a quartering party commander (usually the HHC commander); an S4 representative; company representatives to include supply and NBC personnel; and communication, security, and medical personnel.

(2) The quartering party commander indicates the location of major subordinate units on the ground, formulates a plan to receive and guide units from the RP to their areas, and selects exact locations for the battalion command and administrative installations based on the general location of these areas selected by the S3.

(3) Company representatives select locations for company headquarters, platoons, feeding areas (kitchen areas, if mess is under company control), and latrines.

(4) Communications personnel install equipment that will ensure immediate control of units as units arrive in their assigned areas.

(5) Medical personnel advise other quartering party personnel on sanitation measures and select a site for the aid station.

(6) Based on the order of march, a plan is prepared to guide each unit over a designated route. This route begins at the RP and extends to the unit's new area. Guides must understand and must rehearse the plan. This prevents congestion or delay near the RP. The actual dispatch of the quartering party can follow the issuance of the movement order.

3-2. TIME-DISTANCE TERMS AND FACTORS

The planner must understand march terms to develop detailed movement plans. These terms, along with basic factors of distance, rate, and time, are transformed into movement formulas. Then, formulas are applied to known data to obtain information needed to prepare a time schedule. The time schedule is used to regulate departures and arrivals of march elements.

a. **Time-Distance Relationship.** Relationships between time and distance are the basis for march planning. The planner must determine how far the column is to travel *(distance)* and how long it will take to make the move *(time).* He must know the space *(length of column)* the column will occupy on the route. He must also include in his computation the safety factor of distance *(road gap)* or time *(time gap)* that must separate march columns and their elements. Each term used for distance has its corresponding term for time. The length of a column in kilometers has an equivalent pass time in minutes; the road distance in kilometers or miles has a corresponding time-distance (Figure 3-1).

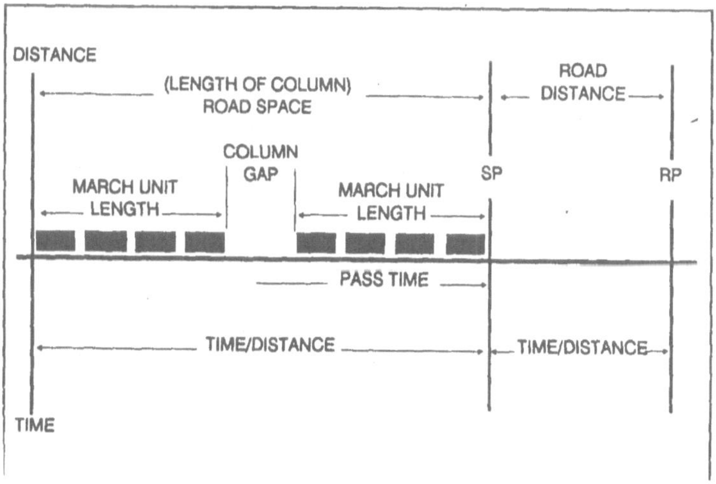

AVERAGE RATES OF MARCH FOR:	KMPH ON ROADS DAY	NIGHT	CROSS-COUNTRY DAY	NIGHT	KM DAYS
FOOT TROOPS	4	3.2	2.4	1.6	20-32
TRUCKS, GENERAL	40	40 (LIGHTS) 16 (BLACKOUT)	12	8	280
TRACKED VEHICLES	24	24 (LIGHTS) 16 (BLACKOUT)	16	8	240
TRUCK-DRAWN ARTILLERY	40	40 (LIGHTS) 16 (BLACKOUT)	12	8	280
TRACTOR-DRAWN ARTILLERY	32	32 (LIGHTS) 16 (BLACKOUT)	16	8	240

Figure 3-1. Time-distance relationships.

b. **Checkpoint.** Checkpoints on a route are those points used for reference in providing instructions, or places where timing might be a critical factor. The route reconnaissance report or a map study should provide the march planner with information to designate checkpoints along the route of march and distances from one checkpoint to another. Once identified, guides and signs are usually sufficient. The commander may want to be present at the passing of some checkpoints. Start points and release points are checkpoints that are always designated.

c. **Start Point.** SPs provide all units of a march column a common point for starting their movement. When units use more than one route, each route has a SP. The SP is a place along the route of march that is easily recognizable on the map and on the ground such as a road intersection. The SP should not be in a defile, on a hill, or at a sharp curve in the road. It should be far enough away from the assembly areas to allow units to be organized and moving at the prescribed speed and interval when the SP is reached. No element of a march column should be required to march to the rear or through another unit in order to reach the SP.

d. **Release Point.** The RP provides all units of the march column a common point for reverting to control of their parent unit. The RP should be on the route of march and easily recognizable on the map and on the ground. Units do not stay at the RP. Guides meet units as they arrive at the RP and lead them to the new areas. Multiple routes and cross-country movement to assembly areas enable units to disperse rapidly. In selecting the RP, units should avoid hills, defiles, and sharp curves. Units should not be required to countermarch or pass through another unit to reach its new position.

e. **Distance Factors-Foot or Motor Marches. The** battalion is normally organized into company-size march units to facilitate control and to maintain unit integrity. The normal march formation is a column of twos. Normal distance is 2 to 5 meters between soldiers (1 to 3 meters at night), 50 meters between platoons (25 meters at night), and 100 meters between companies (50 meters at night).

(1) Vehicle/individual distance is the space between two consecutive vehicles/individuals of an organized element of a column.

(2) Column gap is space between two organized elements following each other on the same route. It can be calculated in units of length or in units of time as measured from the rear of one element to the front of the following element.

(3) Traffic density is the average number of vehicles that occupy 1 mile or 1 kilometer of road space, expressed in vehicles per mile (vpm) or vehicles per kilometer (vpk).

(4) Length of a column is the length of roadway occupied by a column to include gaps in the column measured from front to rear, inclusive.

(5) Road gap is the distance between two march elements—it is the length aspect of column gap. Since it is more significant when the column is moving than when it is halted, road gap becomes a factor of time rather than distance.

f. **Rate Factors.** Speed indicates actual rate of speed of a vehicle or foot column at a given moment as shown on the speedometer (in kph or mph).

(1) Pace is the regulated speed of a column or element that is established by the lead vehicle or an individual in the lead element to maintain the prescribed average speed. For a foot march, the normal pace is 30 inches at a cadence of 106 steps per minute.

(2) The commander considers all of the factors that affect marches and selects a rate that will place his unit at its destination in the shortest time and combat-ready condition. The unit SOP usually states the rate for marches on roads and cross-country, over normal terrain, and day or night. (See Appendix A.) The column commander modifies this rate to suit his needs, which varies greatly in mountain, jungle, desert, or arctic areas. Rates of march usually prescribed for normal terrain are as follows:

	Roads (kph)	Cross-Country (kph)
Day	4.0	2.4
Limited Visibility	3.2	1.6

Marches conducted in mountains, jungle, desert, or northern areas, are characterized by the following:

- Physical effort of individual soldier increases.
- Soldier's load should decrease.
- Potential for injury increases.

g. **Time Factors.** The measurement of time includes the total time needed for the unit to complete the march or to pass a leader point along the designated route. Time is usually measured in minutes or hours.

(1) Arrival time is when the head of the column arrives at a designated point or line.

(2) Clearance time is when the tail of a column passes a designated point or line.

(3) Completion time is when the tail of a column passes the release point.

(4) Pass time (PST) is actual time between the moment the first element passes a given point and the moment the last element passes the same point.

(5) Road clearance time is the total time a column requires to travel over and clear a section of road. Road clearance time equals time distance plus column pass time.

(6) Time distance (TDIS) is time required to move from one point to another at a given rate of march. It normally represents the movement of the head of the column from the start point to the release point.

(7) Time gap is time measured between rear and front of successive elements as they move past a given point. It is the time aspect of column gap or the conversion of road gap to time. There are no prescribed time gaps. Gaps depend on the size of serials and march units, time available for movement, and tactics required for protection against air and nuclear attack.

3-3. MARCH COMPUTATIONS

Before issuing the OPORD, the S3 must compute required time and space measurements to prepare a road movement table.

a. Distance, rate, and time are the factors for movement computations. If two of these factors are known, the third can easily be determined by dividing or multiplying one of the known factors by the other.

- Rate is determined by dividing distance by time:

$$R = \frac{D}{T}$$

- Distance is found by multiplying rate by time:

$$D = R \times T$$

- Time is calculated by dividing distance by rate:

$$T = \frac{D}{R}$$

The march planner must determine pass time, time distance, arrival time, and completion time.

b. Time distance (TDIS) is determined by dividing distance to be traveled by the rate of march. TDIS does not include time for long delays or extended scheduled halts.

$$\text{TDIS (hours)} = \frac{\text{Distance (km)}}{\text{Rate (kph)}}$$

A time distance table is a valuable source for the march planner. It provides a listing of factors used to calculate time required to travel certain distances at specified speeds, either by vehicle or on foot. Travel rates are expressed in vehicle/foot speeds and corresponding rates of march. Travel factors are derived from rate of march, which includes time for short periodic halts and other minor delays that could occur.

c. The length of column (LGTHCOLM) is used to determine the pass time (PST) of a column and consists of two parts: the space occupied by the soldier alone (including the distance between soldiers) and the sum of the distances between the elements of the foot column (column gap). The total length of column is the sum of the two parts.

(1) *Foot elements.* The length of a column of soldiers only is determined by multiplying the number of soldiers by the appropriate factor selected from Table 3-1. This does not include distances between units.

TROOPS LGTHCOLM * = (NUMBER OF SOLDIER'S X FACTOR) + COLUMN GAPS (BETWEEN UNITS)

FACTOR TABLE

FORMATION	2m/MAN	5m/MAN
SINGLE FILE	2.4	5.4
COLUMN OF TWOs	1.2	2.7

* LENGTH OF COLUMN IN METERS

Table 3-1. Determination of length of column (soldiers only).

(2) *Total distance.* The total distance (column gap) between units is obtained as follows:

(a) Determine the number of serial distances (total serials minus one).

(b) Determine the number of march unit distances (total march units minus one, minus the number of serial distances).

(c) Multiply the number of distances obtained by the length in meters between respective units.

(d) Add the results.

EXAMPLE

A battalion foot column is organized into 12 platoon-size march units and three company-size serials. Required: total column gap distances when there are 100 meters between serials and 50 meters between march units.

Serial Distances	=	(3-1) x 100	=	200
March Unit Distances	=	(12-1-2) x 50	=	450
Total Column Gap			**=**	**650 meters**

d. Pass time (PST) for a serial is determined by adding march unit pass times to include time gaps between march units. For foot columns, the PST is determined by applying the following formula:

PST (minutes) = LGTHCOLM x FACTOR (for appropriate rate of march).

EXAMPLE

.0150 for 4.0 kph
.0187 for 3.2 kph
.0250 for 2.4 kph
.0375 for 1.6 kph

Determine the pass time of a unit whose length of column is 1,500 meters and is marching at a rate of 4 kph.

PST(min) = 1,500 x .0150 (factor for 4.0 kph) = **22.5 min**

e. In march planning, the RP is normally designated as the terminal point of movement. Arrival time (AT) at the RP is determined by adding time distance and any long or scheduled halts to the SP time.

EXAMPLE

Determine arrival time for a serial with a SP time of 0800 hours, time distance of 6 hours and 45 minutes, and scheduled halt of 1 hour.

	Hours	Minutes
SP Time	8	00
TDIS	6	45
Scheduled Halt	1	00
	15	45

Arrival time is 1545 hours.

f. Completion time is calculated by adding pass time to arrival time or by adding to the SP time the time distance, pass time, and any long or scheduled halts (other than normal hourly halts).

EXAMPLE ONE

Determine completion time for a serial with an arrival time of 1545 hours and a pass time of 41 minutes.

	Hours	Minutes
AT	15	45
PST	=	41
	15	86

Completion time is 1626 hours.

Note: Convert 86 minutes to 1 hour and 26 minutes, then add to 1500 hours.

EXAMPLE TWO

Determine completion time for a serial with a start point time of 0800 hours, time distance of 6 hours and 45 minutes, a pass time of 41 minutes, and a scheduled halt of 1 hour.

	Hours	Minutes
SP Time	8	00
TDIS	6	45
PST	–	41
Scheduled Halt	1	00
	15	86

Completion time is 1626 hours.

g. Based upon previous movements by a unit, data are accumulated to facilitate march planning. Such data include approximate pass time for various elements of the battalion. The S3 can use these data rather than computing them each time a march is scheduled. Such experience tables serve to reduce the time required to complete the computation phase of march planning. Information that is appropriate to the unit SOP should be integrated.

3-4. ROAD MOVEMENT TABLE

A road movement table is usually an annex to a movement order. It is a convenient means of transmitting time schedules and other essential details of the move to subordinate units. The table is particularly useful in preventing complication of the OPORD or in creating an unusually long OPORD.

a. The road movement table consist of two parts (Figure 3-2):

(1) Data paragraphs that include general information common to two or more march elements; and a list of serials or march units along with all other required information, arranged in tabular form.

(2) Data transferred from the road movement graph. Of particular importance to the march planner are the times at which serials/march units arrive at and clear critical points.

b. Other information on the road movement table includes serial or march unit number, date of move, units involved, number of vehicles, load class of heaviest vehicle, routes to be used, and a remarks section for details not explained elsewhere.

(See FM 55-30 for a detailed discussion on the preparation and use of road movement tables.)

(CLASSIFICATION)

Annex B - "Movement Table" to Operation Order for Movement No.
Map:

Copy No.
Issuing HQ
Place of Issue
Date-Time Group of Signature
Message Reference No.

General Data:

1. Average speed:
2. Traffic density:
3. Halts:
4. Routes (i.e. between start points and release points):
5. Critical points:
   (a) Start points.
   (b) Release points.
   (c) Other critical points.
6. Main routes to start points:
7. Main routes from release points:

Serial or Movement Number	Date	Unit/Formation	Number of Vehicles	Load Class of Heaviest Vehicles	From	To	Route	Route to Start Point	Critical Points			Route from Release Point	Remarks
									Ref.	Due (hrs.)	Clear (hrs.)		
(a)	(b)	(c)	(d)	(e)	(f)	(g)	(h)	(i)	(j)	(k)	(l)	(m)	(n)

Acknowledge:
Distribution:
Authentication:

(CLASSIFICATION)

Figure 3-2. Example of road movement table.

3-5. STRIP MAP

A strip map is a sketch of the route of march and is normally included as an annex to the movement order (Figure 3-3). Enough strip maps should be reproduced to give to key personnel, including vehicle commanders and road guides. The amount of detail depends upon the purpose of the strip map and the unit level at which it is prepared. A strip map should contain the SP and RP, restrictions, and critical points with the distance between them.

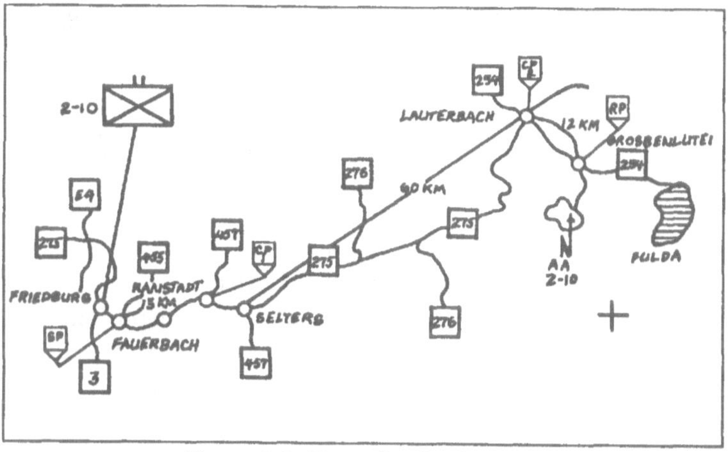

Figure 3-3. Example of strip map.

CHAPTER 4

Execution of FOOT MARCHES

The execution phase of foot marches is a continuation of the planning phase. The troop-leading procedure provides a framework for feedback to the unit commander or leader. Proper planning ensures a smooth transition by setting up proper task organization, security measures, and a flexable command and control system. Also, the successful execution is enhanced by being able to adapt the plan to changing situations.

Section I
CONDUCT OF THE MARCH

Proper execution of the march depends upon establishment of an effective organization and security measures, standardized communication means, and contingencies for reaction to enemy contact. These ensure the foot march is flexible to changing conditions and responsible to the needs of the commander.

4-1. ORGANIZATION FOR THE MARCH

Execution depends upon the establishment of the proper organization and the accomplishment of critical tasks. These ensure flexibility to changing conditions and responsiveness to the needs of the commander. A command is organized into march units and further divided into march serials or march columns (Figure 4-1).

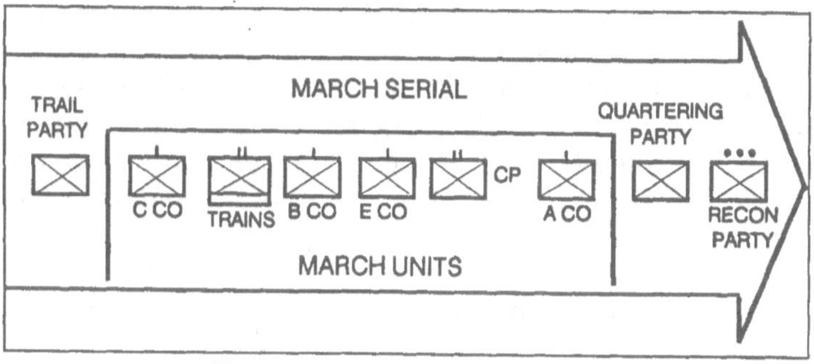

Figure 4-1. March Organizations.

a. **Reconnaissance Party.**

(1) The reconnaissance party reconnoiters the route to determine travel time, capacity of underpasses and bridges, and locations of culverts, ferries, and fords; and to identify critical points and obstacles. Route reconnaissance can be conducted to confirm and supplement data from map studies, higher headquarters, and air reconnaissance. The amount of detail is often related to speed of movement.

(2) Instructions to the party should state the type and extent of information required, and the time and place the report is to be submitted.

b. **Main Body.**

(1) Before starting a march, each march unit of a serial reconnoiters its route to the SP and determines the exact time for reaching it. The movement order states the time that the serial must arrive and clear its SP. The serial commander then determines and announces the times for march units of his serial.

(2) Arrival time at the SP is critical. Each march unit must arrive at and clear the SP on time, otherwise movement of other elements may be delayed. Each leader reconnoiters the route

from his position to the SP to help him decide when his unit must move to meet its SP time.

(3) During the movement, march units move at the constant speed designated in the order, maintaining proper vehicle/foot interval and column gap. Erratic increases and decreases in speed, particularly at hills, create an accordion or whipping effect. This could force tail elements to move at increased and unsafe speeds to keep up with the head of the column. If the march unit is behind schedule, it uses only the designated catch-up speed. March units report crossing each control point as directed by the march order. During the move, air and ground security must be maintained.

c. **Trail Party.**

(1) The trail party is the last march unit in a battalion serial. It is usually led by the BMO and consists of elements of the maintenance platoon and medical personnel. The trail party recovers disabled vehicles and stragglers. If a disabled vehicle cannot be repaired or towed, the vehicle and crewmembers are moved off the road into a secure area. The drivers/crewmembers are left with the vehicle, along with food and water. When a vehicle(s) is left behind, the BMO calls in the location and tells the battalion S4 why it was left.

(2) Medical personnel attached to the trail party compose the personnel evacuation section. This section is responsible for recovering stragglers from the march column that require medical care. Company medical aidmen try to maintain march unit discipline by treating casualties within their ability, but they must not allow stragglers to delay them or the progress of the foot march. The trail party's personnel evacuation section must have troop-carrying assets to pickup and treat stragglers from the various march units within the battalion serial. Once the trail party picks up a straggler, the S1 should be notified to maintain accountability.

(3) When the trail party completes the road march, the battalion's first priority is to recover vehicles left behind and to return stragglers to parent units. A tactical road march is not complete until all march units, vehicles, and personnel are accounted for.

4-2. SECURITY

During the march, the companies maintain security through observation, weapon orientation, dispersion, and camouflage. Commanders assign sectors of observation to their troops so that there is 360-degree observation. Main weapons throughout the column are oriented on specific sectors. The first elements cover the front, following elements cover alternate flanks, and the last element covers the rear.

a. Planning for and implementing air defense security measures are imperative to reduce the battalion's vulnerability to enemy air attack. The battalion commander must be able to integrate into his fire plan the ADA assets allocated to him. Furthermore, he must ensure that all passive and active air defense measures that could be implemented at company level are well planned and used.

(1) *Passive measures* include the use of concealed routes and assembly areas, movement on secure routes, night marches, increased intervals between elements of the columns, and dispersion when under attack.

(2) *Active measures* include the use of organic and attached weapons in accordance with the unit air defense plan during marches.

NOTE: Each vehicle in a motor march has an air guard to provide air security. Specific vehicles may be designated as air guard vehicles performing air rather than ground observation.

b. Scheduled halts preplanned along the march route for maintenance and rest, or to follow higher-echelon movement orders. They should be located on defensible, covered, and concealed terrain. During scheduled halts, vehicles/soldiers pull to the side of the road while maintaining march dispersion. Local security, to include at least one OP for each platoon, is immediately established, and drivers perform during-operation maintenance checks. OPs should not be established outside the range of small arms and should be readily retrievable so that the unit is ready to move at a moment's notice.

c. Unscheduled halts and actions may be caused by unforeseen developments such as obstacles, traffic congestion, or equipment failure. If a halt is necessary, the march column's first priority is to establish security and to have each unit form a hasty perimeter.

d. Obstacles that are reported by the reconnaissance party should be bypassed, if possible. If they cannot be bypassed, the lead march unit assumes a hasty defense to cover and overwatch while engineers remove the obstacle. As the lead march unit removes the obstacle, the other march units move slower or move off the road to closely monitor the battalion command net.

4-3. REACTION TO ENEMY CONTACT

If the battalion is attacked by indirect fire during the road march, the unit in contact continues to move. The remainder of the battalion attempts to bypass the impact area (Figure 4-2).

a. If the battalion is attacked by hostile aircraft during the march, the march unit that is attacked assumes a quick defensive posture or perimeter (whichever is best for the terrain) and immediately engages the attacking aircraft with all available weapons. The rest of the battalion moves to covered and concealed areas.

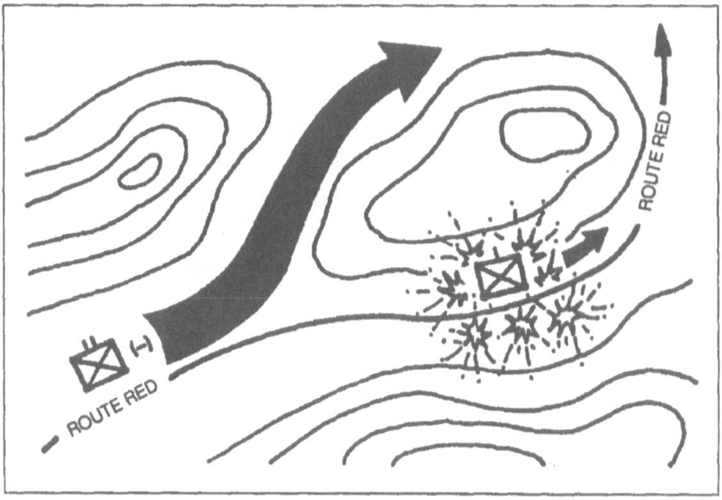

Figure 4-2. Battalion actions under indirect fire.

b. Ambushes are fought through without delay. If the battalion is ambushed, the march unit in the kill zone increases speed, fights through, and reports the ambush. The battalion commander may order a march unit already beyond the kill zone to return to the ambush site (Figure 4-3). The march unit conducts a hasty attack to clear it of enemy or establishes a blocking position on the far side of the kill zone while a following march unit conducts the hasty attack. Follow-on march units may also be ordered to aid in extracting the ambushed march unit from the kill zone either through assault by fire or by direct assault on the ambush position.

c. Disabled vehicles must not obstruct traffic. They are moved off the road, and their status is reported immediately. Security is established and guides are posted to direct traffic. If the operator repairs the vehicle, the vehicle rejoins the rear of the column. If the operator cannot repair the vehicle, trail party maintenance elements pick it up.

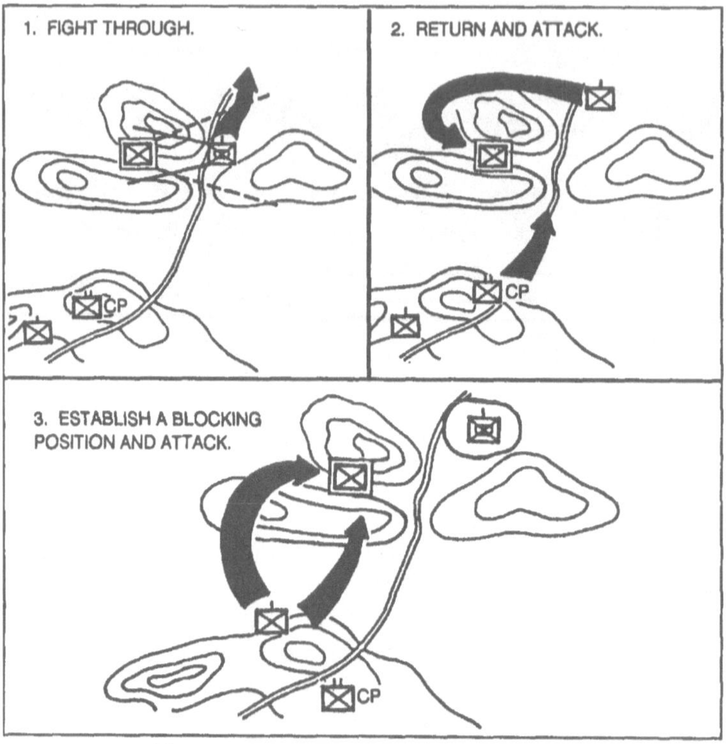

Figure 4-3. Actions on the ambush site.

4-4. COMMUNICATIONS

Messengers are the primary means of communication during tactical marches, but visual signals and road guides can also be used. Since the enemy has good direction-finding equipment, radio is used only in emergencies and when no other means of communication can be used.

a. Road guides can also be used to pass messages from one march unit to a following march unit. Because of the need to stay off the radio, road guides are needed to control the speed of march units and their intervals. When used, they should be given enough information to control movement. The amount of information given depends upon the friendly and enemy situation.

b. The information that messengers or road guides can give to other passing units includes:

- A strip map update.
- The number, sequence, identification, and composition of march units.
- Expected arrival time and clearance time for march units passing the positions of the guides.
- Recognition signals.
- How guides are to be positioned, who will pick them up, and when.
- Instructions for linking up with the parent unit upon completion of road guide duties.
- Special instructions for the guides to pass on to march unit commanders to include details of the route and any changes.

c. Visual communication means include flashlights, lanterns, luminous markers, panel sets, flags, pyrotechnics, smoke, and arm-and-hand signals. The use and meaning of visual signals are standardized to avoid misunderstandings. The road movement order contains instructions for use in specific situations.

d. Sound communications include voice, whistles, horns, and gongs, which can be used to assemble troops upon completion of halts or to warn troops of an enemy air or chemical attack. Sound signals normally are standardized and can be included in the unit SOP if their use is routine.

4-5. HALTS

During a march, halts are enforced routinely to rest personnel and adjust equipment. They are regulated by SOP or by the road movement order. Unit commanders are promptly notified of the time and approximate length of unscheduled halts.

a. Day marches should end early to provide troops with rest and time to prepare for the next day's activities. Midday heat or enemy action can require the unit to make long daylight halts or night marches. At long halts, each unit moves to a planned location near the route of march.

b. Under normal conditions, a 15-minute halt provides rest for soldiers after the first 45 minutes of marching. Following the first halt, a 10-minute halt is enforced for every 50 minutes of marching. Variations of time schedules are required if a halt time occurs when passing through a built-up area, or when cover and concealment are required but not available. On extended or long marches, variations in the time schedule should be kept to a minimum. Units establish observation posts and patrols for security during halts.

c. All units in the column should be halted at the same time. At the halt signal, troops should move to the sides of the road while staying close to their unit. If the situation permits, soldiers should remove or loosen their gear and sit or lie down with their feet elevated for optimum relaxation. Commanders inspect soldiers and equipment during halts, and aidmen administer medical treatment if needed.

d. Soldiers who relieve themselves should dig individual cat-hole latrines and cover them immediately after use. They can dig straddle trenches during noon halts and while occupying bivouac or assembly areas.

4-6. ACCORDION EFFECT

An accordion effect in marching is caused by changes in the rate of march and most often occurs as lead elements of a unit ascend or descend terrain, or pass through critical points along the march route. A change in the rate of march increases as it passes down the column, so that the rear elements must either double time to maintain the distance or be left far behind. Thus, a minor change of rate at the head of the column becomes magnified by the time it reaches the tail of the column. The best method for reducing the accordion effect is for lead elements to slow their rate of march after passing obstacles to permit rearward elements to maintain distances without running. The accordion effect can also be reduced by lengthening or shortening the step, or by taking up the slack to maintain the prescribed distance between soldiers.

4-7. DISTANCES BETWEEN UNITS

Terrain, weather, and the enemy situation influence distances between soldiers and units during a march. Distances should sustain march efficiency and reduce the effects of accordion action, which usually results when marching over hills or difficult terrain. Normal distances are usually included in the unit SOP as a guide to commanders.

a. **Distances Between Men.** When soldiers march on roads in the daytime, the distances between them vary from 2 to 5 meters to provide dispersion and space for marching comfort. A distance in excess of 5 meters increases the length of the column and hinders control. At nighttime, the distances should be reduced to 1 to 3 meters between soldiers to assist in maintaining contact and facilitating control. The tactical situation could require changes to these distances—for example, if the march route is within range of enemy artillery fire, the maximum distances should be used.

b. **Distances Between Units.** Normally, distances are 100 meters between companies and 50 meters between platoons. During reduced visibility, the distances can be decreased to 50 meters between companies and 25 meters between platoons to facilitate control. In the daytime or when the column is marching within the range of enemy artillery, the distances should be increased IAW the situation and the amount of control required. These distances permit vehicles to pass the column.

4-8. NUCLEAR, BIOLOGICAL, CHEMICAL

The battalion commander must include in his plan road marches and assembly areas on the integrated battlefield. He must also consider chemical, biological, and nuclear attacks. These planning considerations should include: properly distributing NBC protective and decontamination materials and ensuring locations are known, establishing and maintaining proper MOPP level, and avoiding contaminated areas.

a. If a unit must cross a chemically contaminated area—

- Use MOPP4.
- Avoid moving through underbrush.
- Stay on hard-surface roads.
- Ensure soldiers are issued NAAK-MK 1 auto injectors.
- Cover as much equipment as possible.
- Avoid vehicle tracking (stagger vehicles in the column) to reduce secondary contamination from dust or debris created by vehicles.
- Avoid low areas.
- Avoid moving early or late in the day.
- Decrease speed to reduce dust or mud.
- Increase vehicle intervals.
- Scrape dirt-road surfaces with a dozer to clear the road of surface contamination.

b. If a unit must cross a nuclear contaminated area—

- Wear BDUs with sleeves rolled down and the top collar buttoned, along with a scarf or handkerchief over the nose and mouth.
- Avoid disturbing road dust.
- Ensure that the unit's radiac equipment is used by march units.

- Dampen hardtop roads before traveling.
- Dampen dirt roads to prevent disturbing contaminated dust.
- Use dosimeters throughout the march to measure the total dose soldiers receive.

c. Detailed planning for the decontamination of march units must be accomplished to reduce losses and excessive delays. Every effort should be made to avoid contaminated areas.

Section II
PERSONNEL DUTIES

This section discusses the minimum duties of individuals at the company level.

4-9. COMPANY COMMANDER

The company commander has a variety of duties before, during, and after the march.

a. **Before the March.** The company commander issues the warning order in time to allow the subordinate leaders and their soldiers to prepare for the march. He conducts reconnaissance of the route to the SP to determine the length of time it will take the company to arrive there from its present area. (If the company is marching alone, the commander selects the route of march and directs a reconnaissance.) Then he prepares a march plan and issues the road movement order. The order is based on the order received from battalion; however, if the company is marching alone, the order is prepared by the company. The order can include, but is not limited to, the following:

- Enemy situation.
- Friendly situation.
- Purpose of the march.
- Units participating.
- March objective.

- Order of march.
- Rate of march.
- Route of march.
- Start point and time.
- Release point.
- Instructions concerning march, water, light, and communication discipline.
- Formation time, based on time distance to start point.
- Uniform and equipment to include water and rations to be carried.
- Location of march collecting posts.
- Designation of personnel for straggler identification and control.
- Command and control.

NOTE: See Appendix B for an example of a battalion road movement order.

The company commander issues instructions concerning personnel who will not march with the unit but will either be left behind or transported to the destination. He supervises preparation for the march. He forms his company at the scheduled time and conducts an inspection to ensure that uniforms and equipment are as prescribed and are worn properly.

b. **During the March.** The company commander usually marches at the head of the company to the SP, after which he moves to a point in the column where he can effectively supervise the march. During the march, he supervises the rate of movement, company formation, appearance of troops, and adjustment of packs, weapons, and other equipment. The commander maintains control by using voice, messenger, and radio communications. Periodically, he checks the march rate and distances between soldiers and platoons to reduce the accordion effect (see paragraph 4-6). He checks the number and condition of stragglers, and ensures that they have been accounted for by the chain of command.

c. **During Halts.** The company commander ensures foot inspections, water discipline, sanitation, safety, and adjustment of loads take place throughout the the company.

d. **After the March.** He ensures that movement of platoons into their respective assembly areas is rapid and without delay. He visits the platoon areas and supervises the platoon leaders in their inspections and execution of other duties.

4-10. COMPANY EXECUTIVE OFFICER

The company executive officer assists the company commander as directed. When the commander is not marching at the head of the column, the company executive officer leads the unit over the designated route at the prescribed rate of march. He supervises the pace setter, assists in supervision of the platoons during halts, and resumes the march according to the time schedule. He posts and supervises company traffic guards and guides. Upon completion of the march, he supervises the establishment of the company command post (CP) and latrine facilities.

4-11. COMPANY FIRST SERGEANT

The company first sergeant assists the commander as directed and ensures the unit NCOs perform all inspections in the conduct of the foot march. He helps establish and maintain march unit formation and discipline to include pace, intervals between soldiers, noise and light discipline, and water and rest discipline.

4-12. PLATOON LEADER

The platoon leader has a variety of duties before, during, and after the march.

a. **Before the March.** The platoon leader informs his platoon of information contained in the road movement order to ensure adequate and timely preparations. During the company formation, just before the march, the platoon leader inspects the wearing and adjustment of clothing and equipment. He supervises his squad leaders in the performance of their duties.

b. **During the March.** As the platoon marches to the SP, the platoon leader ensures that the prescribed distances within his platoon and between his and the platoon ahead are maintained. He checks compliance of the announced restrictions such as water and march discipline. After leading his platoon across the SP, the platoon leader moves to a point in the column where he can effectively control his unit. As the column approaches the RP, the platoon leader moves to the head of the platoon column to lead the platoon into the assembly area.

c. **At Halts.** During halts, the platoon leader directs soldiers to clear the road and to relax by sitting or lying down and by loosening their equipment. He checks the physical condition of his soldiers, and enforces water and food discipline and field sanitation measures. He supervises the formation of the platoon about one minute before the march so it will not be delayed.

d. **After the March.** Upon crossing the RP, the platoon leader ensures that the platoon moves promptly to its assigned area. In the assembly area, he disperses his unit and finds cover and concealment. He ensures that soldiers obtain food, water, shelter, and rest. In a tactical situation, the platoon leader ensures his soldiers are ready to accomplish their mission. He supervises his squad leaders in their duties, while he concentrates on foot inspections. He ensures medical attention where needed.

4-13. PLATOON SERGEANT

Platoon sergeants assist the platoon leader in the conduct of the foot march as directed. He supervises the inspection of soldiers during halts and controls straggling and dropouts by examining march casualties with the medical aidman before recommending to the platoon leader that the casualties be allowed to fall out and take wheel transportation. The platoon sergeant coordinates with the company supply representative for replenishment of water, rations, and medical supplies at rest stops and halts.

4-14. SQUAD LEADER

The squad leader has a variety of duties before, during, and after the march.

a. **Before the March.** The duties of the squad leader are similar to those of the platoon leader and platoon sergeant in relation to his platoon. Upon receipt of the warning order, the squad leader provides detailed instructions and individual attention. He inspects boots for serviceability, proper fit, and comfort. The squad leader inspects each soldier's socks for cleanliness and holes or for badly mended sections that could cause blisters (see Appendix C). He directs each soldier to carry an extra pair of socks for use during and after the march. He can prepare a kit containing foot powder, gauze, adhesive tape, and other appropriate items for use by his squad during the march. The squad leader also inspects soldiers for proper equipment, adjustment of equipment, and canteens full of water.

b. **During the March.** The squad leader marches at a location within the formation where he can best control the squad, maintaining the prescribed distance from the squad ahead and periodically checking soldiers for proper distance and rate of march. He assists the platoon leader in the enforcement of march discipline, other march control measures, and accountability of all personnel and equipment.

c. **At Halts.** The squad leader assists the platoon leader in the accomplishment of his duties by moving his squad off to the-side of the road to allow soldiers to relax. He can shift crew-served weapons and other heavy loads from tired soldiers to those who are rested. He inspects the physical condition of his soldiers. The squad leader is the key to the success of command supervised drinking. Soldiers must be encouraged to drink water at every halt and during the march to maintain proper levels of hydration.

d. **After the March.** The squad leader leads his soldiers to the squad's sector of the assembly area and disperses them, using all available cover and concealment. He conducts a foot inspection and reports the physical condition of his soldiers to the platoon leader or platoon sergeant. He prepares to continue the mission and informs soldiers of details for mission accomplishment.

4-15. COMPANY MEDICAL AIDMAN

The company medical aidman (if attached) assists the platoon leadership in the assessment and treatment of march casualties. He advises the chain of command on evacuation and transportation requirements of casualties during the march.

4-16. GUIDES

Guides are used to lead or direct a unit over a planned route and into or out of a selected area. They can be placed at points along the march to control direction. Guides are given detailed instructions of their duties to include critical information of the march and, if prepared, a strip map of the march route.

a. If conditions prevent the posting of guides ahead of the column, guides accompany the lead element and are posted at critical points to give directions to the remainder of the column. Examples of critical points are road and trail junctions, especially those where a new direction is taken.

b. Guides either join the tail of the column as it clears their posts or are picked up by unit transportation following the column. They are used at the RP to lead units to their assigned sector of the assembly area.

4-17. ROAD GUARDS

Road guards are placed about 50 meters to the front and rear of the column to slow or stop oncoming or passing traffic. The column commander places guards at road intersections or other critical points to stop traffic while the column crosses. At such times, the road guards also act as guides. When possible during night marches, road guards use flashlights or lanterns to control traffic.

4-18. PACE SETTER

The pace setter is an experienced soldier, carrying the same load as the majority of the soldiers, and marching from 4 to 10 meters at the head of the column (Figure 4-4).

a. The pace of the column must be governed by the most heavily loaded element. The pace setter's primary duty is to maintain the rate of march ordered by the column commander. He does this by establishing his pace *(length of step)* and cadence *(number of steps per minute)* to obtain the prescribed rate of march.

b. The pace setter should be of medium height so average strides can be taken. Overstriding or understriding tends to quickly tire the leg muscles and affects the efficiency of marching troops. The officer or NCO marching at the head of the column supervises the pace setter to ensure that the pace setter takes average strides and maintains a uniform cadence.

Figure 4-4. Pace setter.

Section III
SPECIAL MARCH OPERATIONS

Special march operations include limited visibility marches and forced marches, which require unique considerations in planning and executing. However, the techniques and procedures appropriate for other operations also apply.

4-19. LIMITED VISIBILITY MARCHES

The battalion commander must always be prepared to maneuver and engage his unit under all conditions.

a. Units must routinely operate during limited visibility. The battalion commander coordinates and controls the movement of his unit, but, as with any mission, disadvantages and advantages exist when operating during limited visibility.

(1) **Disadvantages.**

- Difficulty in navigation.
- Slower rate of speed.
- Difficulty in recognizing checkpoints.
- Difficulty in control of subordinate units.
- Difficulty in maintaining the proper interval between units.

(2) **Advantages.**

- Increased security.
- Tighter formations.
- Less traffic at night.
- Use of surprise and stealth.

b. Battalion SOPs should reflect increased control when movement must be during limited visibility. Items may include:

(1) Assigning colors to march units, which may be used on flashlights for recognition.

(2) Closing the interval between elements of the column.

(3) Increasing use of connecting files and messengers between march units in the serial.

(4) Requiring the dismounting of one soldier during short halts in a motor march to move to the halted vehicle to the front. The soldier informs the driver that the march has begun. This technique ensures constant contact within the column and prevents a break in the column. It also applies to foot marches through the use of connecting files.

c. Limited visibility marches are characterized by tighter formations, difficult control and reconnaissance, and a slower rate of march than day marches.

(1) Control is increased by reducing the distances between soldiers and units. The number of guides can be increased, depending on the suitability of the roads or trails. Also, visual communication means are used such as flashlights, lanterns, luminous markers, and pyrotechnics (consistent with light discipline).

(2) Limited visibility marches require added safety measures to prevent accidents, which should be provided by unit and installation SOPs. The following measures are appropriate:

(a) Use off-road trails or routes to avoid roads used by vehicles.

(b) Place guards to the front and rear of the column and on the flanks when vehicles could approach from that direction. Ensure guards are equipped, marked, and informed of the enemy situation.

(c) If the tactical situation permits, mark moving or static traffic guards, and other key personnel with reflective or luminescent materials such as reflective fabric or tape, vests, caps, mittens, hat bands, and traffic ensembles. Equip guards with appropriate warning devices such as illuminated or reflective signs, red lights, flashlights, traffic control batons, reflective flags, and lanterns.

(d) Warn vehicle operators of the presence of troops on or near the roadway and limit speeds as needed.

(e) Provide for the exclusive use of selected routes by foot soldiers. Enforce safety measures even when troops are assigned exclusive use of routes that are negotiable by wheeled or tracked vehicles.

4-20. FORCED MARCHES

A normal foot march day is 8 hours, for a distance of 32 km at a rate of 4 kph. The maximum distances recommended for forced marches are: 56 km in 24 hours; 96 km in 48 hours; or 128 km in 72 hours. A forced march usually exceeds this distance by increasing the hours marched rather than by increasing the rate of march. However, sometimes the rate of march must be increased to adjust to the situation.

a. Although forced marches may impair the fighting effectiveness of a unit, urgent conditions on the battlefield could require them. Rest periods should be scheduled to avoid marching at the hottest time of day and to ensure the arrival of the unit in combat-ready condition. At this time, commanders should consider increasing the rate of march when soldiers are most rested.

b. Time for a forced march of 52 km (4 km less than the maximum recommended distance) should be scheduled as follows, assuming the march began at first light:

	Hours
First phase: 20 km at 4 kph (daylight, on roads)	5.0
Noon meal and rest period	2.0
Second phase: 20 km at 4 kph (daylight, on roads)	5.0
Supper meal and rest period	6.0
Third phase: 12 km at 3.2 kph (limited visibility, on roads)	3.84
Total time	**21.84**

CHAPTER 5

Soldier's Load Management and Training for FOOT MARCHES

The ability of a soldier to march and fight is directly related to his load. The maximum individual load limit cannot be exceeded as an infantry soldier will not accomplish his mission. Soldiers fight light with only the equipment required for the immediate mission. They receive additional weapon systems and materiel when required. Effective individual fighting loads and minimum approach march loads can only be achieved through safeguarding and transporting portions of the load–commanders must decide to tailor loads that result from risk analysis. Transportation resources must be used to avoid excessive loads on soldiers such as CLOHE at company level and SLOHE at battalion level.

"No soldier should be compelled to walk until he actually enters battle. From that point forward he should carry nothing but what he wears, his ammunition, his rations and his toilet articles. When battle is concluded he should get new uniforms, new everything."

General George Patton, Jr.

Section I
TRANSPORT RESPONSIBILITY

The soldier load concept transfers responsibility for transporting part of the load from the individual soldier to battalion and division staff, and transfers the load from men to CLOHE. The soldier load problem can be reduced by reorganizing existing transport resources, which entails dedicating resources to the roles of CLOHE and SLOHE.

5-1. SIZE OF COMPANY LOADS

Table 5-1 shows planning figures for the required load-handling lift for each infantry company.

CONFLICT/ ENVIRONMENT	CLOHE (COMPANY LEVEL)	SLOHE (BATTALION LEVEL)
LOW INTENSITY		
TEMPERATE CLIMATE	1,850 LB	4,800 LB/400 CUBIC FT
COLD WET CLIMATE	1,850 LB	7,000 LB/600 CUBIC FT
MEDIUM INTENSITY		
TEMPERATE CLIMATE	2,250 LB	6,350 LB/500 CUBIC FT
COLD WET CLIMATE	2,250 LB	8,550 LB/700 CUBIC FT

Table 5-1. Soldiers loads requiring transportation.

a. **Transportation Resources.** The provision of CLOHE at company level and SLOHE at battalion level is the responsibility of the command level that has control of transportation resources for ongoing operations.

b. **Resupply.** Company commanders will more readily make risk judgments if they have operational control of transportation. They can reduce the weight of ammunition, food, and water carried by their men to carry an immediate resupply, which consists of part of their basic load.

5-2. EXPEDIENTS FOR EXTRA TRANSPORTATION

If extra transportation resources are not given to battalions, greater reliance must be placed on–

a. **Extensive use of helicopters to free unit HMMWVs for use as CLOHE and SLOHE.** This might entail allocating one dedicated helicopter to each infantry brigade for logistical support to release three HMMWVs and trailers required for each company for load handling.

b. **Deployment of corps plug transportation assets.** Corps assets could be placed under the operational control of battalions for use as load-handling equipment. Forward deployment of corps transportation assets in the division can release existing HMMWVs and can improve the soldier load-carrying capacity of units, as will direct resupply of forward units by airdrop or steerable parachutes.

c. **Host nation support.** Units should be prepared to use local resources to include conventional vehicles, agricultural tractors, beasts of burden and their handlers, and human porters, which are obtained through host nation support, renting, and capture. Leaders must know the legal parameters of commandeering equipment and animals. The required funding must be provided for renting equipment. Possible host nation resources should be identified in contingency plans. At least one man in each platoon should be designated as a general-purpose driver.

Section II
FACTORS AFFECTING THE SOLDIER'S LOAD

Commanders at all levels must understand the factors affecting the soldier's load and the subsequent capabilities or limitations produced in the unit. The physical limitations of individual soldiers, stress, and the weight of equipment and munitions all affect the soldier's ability to carry his required load. These factors must be carefully analyzed by the commander or leader in the load determination process.

5-3. PHYSICAL LIMITATIONS

The fighting load for a conditioned soldier should not exceed 48 pounds and the approach march load should not exceed 72 pounds. These load weights include all clothing and equipment that are worn and carried.

a. A soldier's ability to react to the enemy is reduced by the burden of his load. Load carrying causes fatigue and lack of agility, placing soldiers at a disadvantage when rapid reaction to the enemy is required. For example, the time a soldier needs to complete an obstacle course is increased from 10 to 15 percent, depending on the configuration of the load, for every 10 pounds of equipment carried. It is likely that a soldier's agility in the assault will be degraded similarly.

b. Speed of movement is as important a factor in causing exhaustion as the weight of the load carried. The chart at Figure 5-1 shows the length of time that work rates can be sustained before soldiers become exhausted and energy expenditure rates for march speeds and loads. A burst rate of energy expenditure of 900 to 1,000 calories per hour can only be sustained for 6 to 10 minutes. Fighting loads must be light so that the bursts of energy available to a soldier are used to move and to fight, rather than to carry more than the minimum fighting equipment.

c. When carrying loads during approach marches, a soldier's speed can cause a rate-of-energy expenditure of over 300 calories per hour and can erode the reserves of energy needed upon enemy contact. March speeds must be reduced when loads are heavier to stay within reasonable energy expenditure rates. Carrying awkward loads and heavy handheld items causes further degradation of march speed and agility. The distance marched in six hours decreases by about 2 km for every 10 pounds carried over 40 pounds. Figure 5-2 shows speeds that are sustainable with given loads, which results in an energy expenditure of 300 calories per hour.

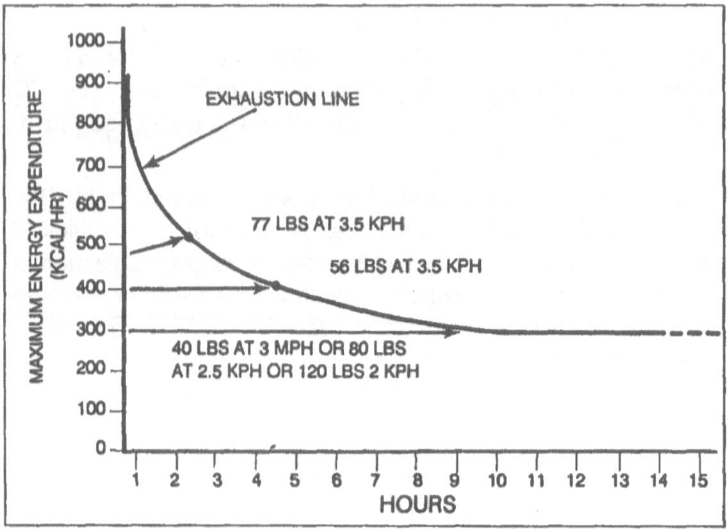

Figure 5-1. Work rate and energy expenditure.

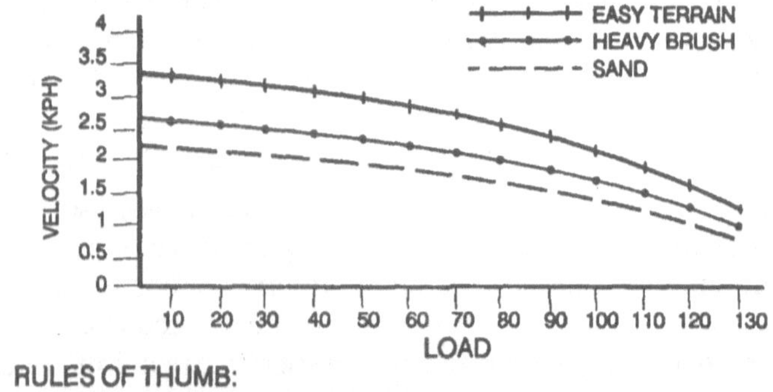

RULES OF THUMB:

1. DISTANCE MARCHED IN 6 HOURS DECREASES BY 2 KILOMETERS FOR EVERY 10 POUNDS CARRIED OVER 40 POUNDS.
2. TIME OVER ASSAULT COURSE INCREASES BY 15 PERCENT FOR EVERY 10 POUNDS CARRIED OVER 40 POUNDS.
3. DISTANCES TRAVELED REDUCED BY HALF OVER AVERAGE GRADIENTS OF 10 PERCENT.

Figure 5-2. March speeds.

5-4. STRESS

Battlefield stress decreases the ability of soldiers to carry their loads. Fear burns up the glycogen in the muscles required to perform physical tasks. This wartime factor is often overlooked in peacetime, but the commander must consider such a factor when establishing the load for each soldier. However, applying strong leadership to produce well-trained, highly motivated soldiers can lessen some of the effects of stress.

5-5. MUNITIONS AND EQUIPMENT

As the modern battlefield becomes more sophisticated, potential enemies develop better protected equipment, which could be presented as fleeting targets. Unless technological breakthroughs occur, increasingly heavy munitions and new types of target acquisition and communications equipment will be required by frontline soldiers to defeat the enemy.

a. In the future, the foot soldier's load can be decreased only by sending him into battle inadequately equipped or by providing some means of load-handling equipment to help him carry required equipment. Weights of selected items are provided in Table 5-2.

WEIGHT IN POUNDS *(every ounce counts)*

BDU	3.8	TROUSERS, WET WEATHER	1.2
DRAWERS, COTTON	.1	RATION, MRE	1.3
HANDKERCHIEF	.1	BAG, WATERPROOF	.8
SOCK, CUSHION SOLE	.2	PAD, SLEEPING	1.3
UNDERSHIRT, COTTON	.3	3 SHELTER HALF, POLES, PEGS, AND ROPE	4.5
TOWEL	.2	CARRIER, SLEEPING BAG	.4
CANTEEN, 1 qt w/WATER	2.8	BOOTS, COMBAT LEATHER	3.3
CANTEEN, 2 qt w/WATER	4.8	JACKET, FIELD	3.3
LINER, PONCHO	1.6	BAG, DUFFLE	3.5
MESS KIT	2.8	CAP, BDU	.3
GLOVES, BARBED WIRE	.4	CASE, SLEEPING BAG	1.5
PARKA, WET WEATHER	1.2	LINER, FIELD JACKET	.7
PONCHO, NYLON	1.3	OVERSHOES	4.2
SHIRT, SLEEPING	.7		

SCARF, WOOL	.4	TELEPHONE, TA-1	1.5
SLEEPING BAG	7.1	E-TOOL, w/CARRIER	2.5
BELT, TROUSERS	.2	ALICE, MEDIUM, w/FRAME	6.3
HELMET, BALLISTIC	3.4	ALICE, LARGE, w/FRAME	6.6
BELT, PISTOL, w/SUSPENDERS AND FIRST-AID POUCH	1.6	AN/PRC-77, w/BATTERY	20.8
TOILET ARTICLES	2.0	M60 SPARE BARREL w/BAG	8.0
WEAPONS:		60-mm MORTAR, M225	14.4
M16	1.6	60-mm SIGHT, M64	2.5
M203	10.0	60-mm BASEPLATE, M-7	14.4
M60 MG	23.3	60-mm BIPOD	13.2
M249 SAW	15.2	81-mm MORTAR, M29	30.0
AMMUNITION:		81-mm SIGHT, M53	6.0
5.56 w/MAG (30 rds)	.9	81-mm NIGHTLIGHT	2.0
7.62 LKD (100 rds)	7.0	81-mm BASEPLATE	25.0
40-mm (ALL TYPES)	.5	81-mm BIPOD	40.0
5.56 LKD (200 rds)	7.6	BINOCULARS	3.2
GRENADE, FRAGMENTATION	1.0	FLASHLIGHT, w/BATTERY	.8
GRENADE, SMOKE	1.0	COMPASS, M2	.3
RD, 60-mm MORTAR, HE	3.5	DRAGON TRACKER	8.1
RD, 81-mm MORTAR, HE	9.3	DRAGON NIGHTSIGHT	34.0
LAW	5.2	AN/PVS-5 NVG	1.9
MINE, M21	18.0	AN/PVS-4 NVD	3.9
CLAYMORE, M18	5.0	PISTOL, CAL .45	2.5
DRAGON, MSL	25.3	PROTECTIVE MASK, w/DECON KIT	3.0
AT4	14.0		
FLARE, TRIP	1.0		
BAYONET, w/SCABBARD	1.3		
CASE, SMALL-ARMS	.9		

Table 5-2. Weights of selected items.

b. Unless part of the load is removed from the soldier's back and carried elsewhere, all individual load weights are too heavy. Even if rucksacks are removed, key teams on the battlefield cannot fulfill their roles unless they carry excessively heavy loads. Soldiers who must carry heavy loads restrict the mobility of their units.

c. Overloaded soldiers include the antiarmor teams (individuals carry weights of 111, 101, and 90 pounds), mortar teams (individuals carrying 83 pounds, even after distributing 100 mortar rounds of 3.5 pounds each), fire support teams (carry 92 to 95 pounds), and M60 machine gun teams (carry 78 to 87 pounds). All radio operators equipped with the AN/PRC-77 and KY57 VINSON secure device are also loaded above the maximum recommended combat load (84 pounds). AT4 gunners and low-level voice intercept teams are overloaded as well as Stinger and engineer breaching teams.

Section III
ECHELONING AND LOAD TAILORING

A diagram showing the concept of dividing the total soldier load into combat, sustainment, and contingency loads and the different levels of combat loads (fighting and approach march) is at Figure 5-3.

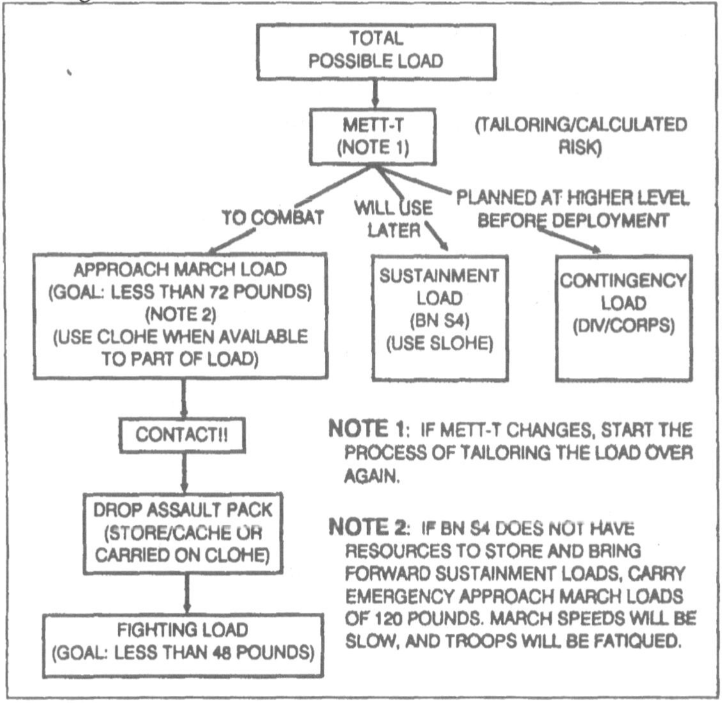

Figure 5-3. Load echelon diagram.

5-6. COMBAT LOAD

The definition of a combat load that is carried by soldiers is as follows: The minimum mission-essential equipment, as determined by the commander responsible for carrying out the mission, required for soldiers to fight and survive immediate combat operations. The combat load is the essential load carried by soldiers in forward subunits or the load that accompanies soldiers other than fighting loads.

a. **Fighting Load.** The fighting load includes bayonet, weapon, clothing, helmet, and LBE, and a reduced amount of ammunition.

(1) For hand-to-hand combat and operations requiring stealth, carrying any load is a disadvantage. Soldiers designated for any mission should carry no more than the weapons and ammunition required to achieve their task; loads carried by assaulting troops should be the minimum.

(2) Unless some form of CLOHE is available, cross-loading machine gun ammunition, mortar rounds, antitank weapons, and radio operator's equipment causes assault loads to be more than the limit of 48 pounds. This weight restricts an individual's ability to move in dynamic operations. Extremely heavy fighting loads must be rearranged so that the excess weight can be redistributed to supporting weapons or can be shed by assaulting troops before contact with the enemy.

b. **Approach March Load.** The approach march load includes clothing, weapon, basic load of ammunition, LBE, small assault pack, or lightly loaded rucksack or poncho roll.

(1) On prolonged dynamic operations, the soldier must carry enough equipment and munitions for fighting and existing until resupply. In offensive operations, soldiers designated as assault troops need equipment to survive during the consolidation phase, in addition to carrying munitions for the assault. A limit of 72 pounds for a soldier load should be enforced.

(2) Normally, the soldier's large rucksack is not part of the approach march load. If the field pack internal frame is issued,

only the small assault pack section is carried—the large section should be kept at battalion level. If the ALICE system is used, either a partly loaded small ALICE should be carried individually with a duffle bag or one large ALICE for each man should be kept at battalion level.

c. **Emergency Approach March Loads.** Circumstances could require soldiers to carry loads heavier than 72 pounds such as approach marches through terrain impassable to vehicles or where ground/air transportation resources are not available. Therefore, larger rucksacks must be carried. These emergency approach march loads can be carried easily by well-conditioned soldiers. When the mission demands that soldiers be employed as porters, loads of up to 120 pounds can be carried for several days over distances of 20 km a day. Although loads of up to 150 pounds are feasible, the soldier could become fatigued or even injured. If possible, contact with the enemy should be avoided since march speeds will be slow.

d. **Shedding Items on Contact.** Rucksacks, assault packs, and other items of the approach march load should be placed in caches before combat, and contact and antiambush drills must include shedding most of the approach march load. However, this procedure makes it difficult for the soldier to retrieve his equipment later in the battle, and it should only be used when transport is unavailable.

5-7. TAILORING COMBAT LOAD TO METT-T

When confronted with the unacceptable loads that soldiers may be required to carry, senior commanders would like to establish firm limits. However, when realistically applying such guidelines, the soldier may not survive and win on the modern battlefield if his munitions and survival items are less effective than his enemy's.

a. Senior leaders must resist the temptation to prescribe the soldier's load. Only the subunit commander, who is assigned a specific mission, has information on all the factors that must be considered to decide the equipment needed for his mission. He must also have the authority to decide what load his soldiers must carry. The subunit commander receives guidance from his superiors, but any standard load imposed on him only inhibits his ability to make the correct judgment.

b. Consideration of risk judgments that must routinely be taken by subunit commanders are at Table 5-3. To use the table, the commander prioritizes the factors in column 1 based on an estimate of the situation. This estimate includes a complete analysis of the unit's mission at the end of the march, the enemy situation, the terrain to be marched on or through, the expected weather, and the physical condition of the troops. After prioritization, the commander decides what items to include in the soldier's load and what items to leave with the sustainment load in the unit trains area.

c. Considerations for this decision-making are included in columns 2,3 and 4. This prioritization allows the commander to tailor the soldier's load based on the mission and to decide what items are to be transported or stored in the sustainment load. The mission analysis also identifies requirements for additional transportation assets.

	RISK ASSESSMENT		
1 FACTOR/QUESTION	2 TAKE	3 LEAVE	4 REMARKS
MISSION			
a. To defeat the enemy in close battle.	Reduced munitions Water	Food Threat protection Environmental protection	Mobility is paramount with 40-pound loads
b. To get there quickly.		Reserve munitions	

Table 5-3. Risk judgments in load planning.

	RISK ASSESSMENT		
1 FACTOR/QUESTION	2 TAKE	3 LEAVE	4 REMARKS
c. To sustain stealth operations independent of resupply.	Water Food Environmental protection Reduced munitions Camouflage	Reserve munitions Threat protection	Maximum loads depend on speed/distance for dynamic operations.
d. To carry maximum combat power.	Munitions Water Threat protection Limited environmental protection	Food	Maximum loads depend on speed/distance for dynamic operations.
e. For static operations.	Basic load and reserve of ammunition Barrier materiel Maximum threat protection Some comfort items to achieve quality rest periods Water Food		
RESUPPLY			
a. Reliability.	Less amounts of all classes of supply		Best solution is for the commander to control his own immediate resupply transport resources.
b. On call.		Reserve munitions	

Table 5-3. (Continued.)

	RISK ASSESSMENT		
1 FACTOR/QUESTION	2 TAKE	3 LEAVE	4 REMARKS
Electronic Warfare	VINSON		Secure communications probably not viable below bde/bn level in light units unless COMSEC is of high priority to achieve mission.
TERRAIN			
Flat, improved road			Terrain may cause an increase of time required to conduct march; resupply cross country may be difficult.
Cross country	Water consumption increased		
Hills, improved road			
WEATHER			Energy must be maintained to fight by control of loads/march speeds.
a. Environmental Survival:			

Table 5-3. (Continued.)

	RISK ASSESSMENT		
1 FACTOR/QUESTION	2 TAKE	3 LEAVE	4 REMARKS
Exposure	Poncho Extra clothing Limited number of sleeping bags		Work rates should be reduced.
Heat exhaustion	Water	Threat protection	
Disease	Water purification tablets Mosquito nets		When in combat, men with excess fat can survive off natural reserves.
b. Sustenance	High-caloric food		Average of four hours quality sleep each day.
c. Quality rest	Sleeping bags/ pads		

Table 5-3. (Continued.)

5-8. ECHELONING THE SOLDIER'S LOAD

When the size of the combat load has been established, needs can be determined for echeloning at battalion-, division-, or force-level trains for other equipment assigned to foot soldiers. Given the approved load limits for individual soldiers (fighting load of 48 pounds; approach march load of 72 pounds), commanders can use the following as a guide for tailoring the equipment and supplies into three echelons. (See Appendix D.)

a. **Combat Load.** Company commanders are responsible for all the equipment included in combat loads.

b. **Sustainment Load.** Battalion S4s are responsible for the safe custody and movement of sustainment loads. This responsibility may be delegated to company supply sergeants. The definition of sustainment load is as follows: equipment that is determined by the commander to be required for sustained operations, which will be brought forward to the soldiers under unit arrangements as required by the commander.

(1) The sustainment load includes individual large rucksacks or A-bags that contain spare clothing and equipment: sleeping bags when not required for survival, limited personal effects, protective items for specific threats (armored vests and chemical protective suits), and a mobile store of unit equipment.

(2) This equipment should be stored in a forward operations base–such as the battalion combat trains area–and pushed forward to front-line soldiers as arranged by the S4. Resupply of ammunition, food, and water is not included in this definition of sustainment load; however, the sustainment load could be pushed forward at the same time as normal resupply.

c. **Resupply.** Combat and sustainment loads contain ammunition, -water, and food to enable units to operate until resupplied, but the amounts carried directly reflect the front-line soldier's confidence in the reliability and frequency of resupply. This concept assumes the normal operation of the battalion resupply system, but transport carrying an immediate company resupply should normally be under the operational control of the company commander.

d. **Contingency Load.** The contingency load is stored under divisional or corps arrangements until required. It consists of all other items of individual and unit equipment not required for ongoing operations such as the soldier's B-bag for extra clothing and TOW missiles in theatres where there is no armored threat.

(1) B-bags should be palletized in company loads, and contingency TO&E equipment should be centralized in battalion packs. Contingency loads are not flown into

deployment areas as part of the initial sorties. When contingency loads arrive in theater, plans should be made at or above division level to store unit contingency equipment. Items of equipment can then be returned to units if the operational situation changes, if the unit is deployed into an area where items of the contingency load are needed, or if the unit is staged through a rear assembly area.

(2) The key to this process it that instructions are issued to soldiers before deployment, listing the items of individual and TO&E equipment that should not accompany them on the initial deployment. Contingency equipment could remain in CONUS, or be stored at a base area in unit packs, or be used as a pool of equipment to be issued as required by the G4.

(3) Provision must also be made for some items of equipment to be back-loaded from battalion to division control upon arrival in theater. This allows units to deploy heavy for maximum flexibility and to add to contingency loads when in theater to fight light.

NOTE: In addition to the LBE, rucksack, A-bags, and B-bags, soldiers leave a C-box of personal equipment at their home station before deployment.

e. **Load Planning.** Units should develop packing lists to include specific deployment options, based on guidance from the chain of command. Once deployment has taken place, authority should be delegated to battalion commanders to send items back to division for inclusion in contingency loads. Company commanders should be authorized to vary the composition of combat and sustainment loads.

5-9. MINIMUM-LOAD CONCEPT

All soldiers, regardless of the threat environment and mission, always carry certain items common to the any mission. These items are the minimum-load configuration (MLC) along with the soldier's assigned weapon system and minimum amount of ammunition. Additions or deletions to the MLC will be based on the unit commander's estimate of the situation.

Section IV
TRAINING

Training in foot marches develops a unit's ability to march to its destination in a condition to accomplish its mission.

5-10. UNIT CHARACTERISTICS

Whether a force is mounted or dismounted, success in combat depends upon troops who can move dismounted cross-country, covering a great distance in the shortest time. The following unit traits are important in achieving this goal: unit discipline, good leadership, teamwork, high morale, endurance, and mental and physical stamina.

5-11. PHYSICAL CONDITIONING

An essential element of training foot soldiers is proper physical and mental conditioning, which develops stamina and endurance to perform required tasks. The best results are obtained from cross-country marches, although physical training and negotiating obstacle courses are also useful. Loads should be light and distances short at the beginning of training, becoming more difficult as training continues. By the end of training, troops should be accustomed to rigorous conditions that are most likely encountered in subsequent operations.

a. Soldiers who are physically fit to APFT standards can carry loads that are 45 percent of their body weight (average 72 pounds) at 4 kph for eight-hour approach marches. The amount of energy expended and discomfort experienced in carrying these loads can be reduced if soldiers have participated in a specialized program of physical conditioning. As a result, much heavier emergency loads can be carried at reduced speeds. Soldiers whose mission is to operate on foot for long periods without resupply can benefit from such training and conditioning.

b. After a 30-day preparatory training period, soldiers can march 12 miles in less than 3 hours loaded to about 60 pounds, when energy expenditure at that rate would cause exhaustion in 2.5 hours for soldiers who have not received special training. A number of considerations should be examined when developing a program for march conditioning.

(1) Aerobic conditioning (running) should not exceed 3 to 5 sessions each week (30 to 45 minutes each session). Excessive aerobic conditioning could be counterproductive to meeting other physical fitness objectives.

(2) Extended marching with loads is the most demanding physical requirement for infantry soldiers. Training programs should include specific progressive load-bearing marches. Progressions should be developed for increasing a load or distance marched but not at the same time. This training meets the requirement for both aerobic and muscular (legs and back) endurance. Progressive extended marches of 5, 10, 15, and 20 miles should be mandatory training and be scheduled regularly.

(3) Infantry physical training programs should include scheduled (two or three times each week) progressive resistance (strength) training to sustain or improve muscular strength. Also, muscular endurance training should be included. This training should include both upper and lower body muscle groups with emphasis on the upper body. Intensive progressive resistance training (three sessions each week in 48-hour intervals) should be mandatory for soldiers who lack adequate muscle mass (lean tissue) of the upper body.

(4) Mandatory elements of any fitness program should be: load-bearing progressive marches; resistance strength training; aerobic training; and anaerobic (speed and agility) training to include negotiating obstacles and confidence courses.

c. The following is a suggested program for physical conditioning.

(1) Train-up program for six weeks consisting of four one-hour daily workouts and one-half day each week to include:

(a) Two upper body exercise periods (push-ups, dips, sit-ups, and chin-ups or pull-ups).

(b) Two lower body exercise periods (sprints, relays, fireman carry, boot dusters, step-ups on benches).

(c) Two marches: one with heavy load and a short distance, one with light load and a longer distance, or both combined with tactical missions.

(d) Two slow-paced distance runs of 3 to 5 miles at 80 percent maximum heart rate *(FM 21-20).*

(e) One light run of two miles at 60 to 80 percent maximum heart rate.

(2) Sustainment program as determined by the commander based upon the seven physical training principles *(FM 21-20).*

(a) Regularity (three to five times each week).

(b) Progression (slow, steady increase of load or distance).

(c) Overload (work until muscles are fatigued).

(d) Variety (circuit training, free weights, calisthenits/isometrics, confidence and obstacles courses).

(e) Balance (flexibility and muscular balance).

(f) Recovery (stress different muscle groups each day).

(g) Specificity (body adapts to a specific demands).

5-12. NUTRITION

Proper nutrition planning can improve soldier performance and reduce exertion required for a given work load. A preload-bearing diet that is high in carbohydrates can add to a soldier's ability to carry his load. It increases muscle glycogen levels as a "buffer" or reserve against exhaustion. Soldiers with high glycogen levels could require less rations for short load-bearing missions than soldiers low in muscle glycogen.

5-13. TACTICAL TRAINING

Units should train regularly with their dummy basic load of ammunition. The execution of a platoon load-carrying test should be included in the ARTEP – for example, carrying an average load of 70 pounds for 12 miles followed by a tactical exercise, followed by an additional 12 miles and tactical exercise to be completed in a period of 24 hours.

5-14. LEADER TRAINING

Improved use of aimed fire and fire discipline decreases the risk of depleting ammunition, allowing commanders to carry reduced loads of ammunition when foot mobility is paramount.

a. **Commanders.** Potential commanders and staff officers must understand the following points:

(1) The composition of loads depends on METT-T factors. Load planning involves the subunit commander, acting under guidance from his superior, in a series of risk judgments that require balancing the physical ability of his soldiers against the risks of not carrying items of clothing, equipment, food, water, or munitions.

(2) Levels of command above company must support the front-line soldier by arranging for items of individual and subunit equipment to be kept in trains or base areas and by providing resources to fulfill the role of CLOHE and SLOHE. Echeloning of loads must be planned so that combat loads, sustainment loads (battalion level), contingency loads (divisional level), and equipment at home base are properly accounted for, safeguarded, and available to the soldier.

b. **Junior Leaders.** Junior leaders should be taught to make risk jugments involved in load planning and movement as well as load discipline.

5-15. SUSTAINMENT (INTEGRATED) TRAINING

Previously learned subjects should be integrate with foot marches to maintain overall proficiency and to inject realism into training. The subjects to be integrated and their required emphasis depend on the area of operations in which the training is conducted. A list of subjects that apply to most disciplines includes:

- Camouflage.
- Reconnaissance and security.
- Map reading.
- Use of the compass and other navigational aids.
- Fire support planning.
- Air defense planning.
- NBC defense planning.
- First aid and personal hygiene.
- Field sanitation.
- Occupation of bivouac and assembly areas.
- Tent pitching/field craft.
- Preparation of individual and small-group rations.
- Evacuation of casualties.
- Care of clothing and equipment.
- React to ambushes.
- Use of arm-and-hand signals.
- Request for indirect fire.

5-16. ENVIRONMENTAL TRAINING

Training should consider the terrain and climate of the area in which the unit will subsequently conduct operations. The training program must include familiarization with special equipment and the application of specialized techniques to tactical principles. Specialized training procedures for desert, jungle, arctic, and mountain areas are found in manuals dealing with those areas of operations.

5-17. MARCH DISCIPLINE

March discipline must be stressed throughout training. Aspects requiring special consideration are maintaining the rate of march and distances between men and units, and ensuring proper timing and use of halts and rest periods. Troops must not drop MRE wrappers or other refuse along the trail, and they must observe prescribed sanitation procedures. At halts any material that could attract the attention of the enemy or could identify a force should be removed or buried. When contact with the enemy is imminent, noise discipline must be observed for required security.

APPENDIX A

UNIT STANDING OPERATING PROCEDURES FOR FOOT MARCHES

A unit SOP presctibes routine methods to be followed in operations. To provide for quick and efficient movement, individual and unit training in preparation for and conduct of all forms of movement is desirable. The unit SOP should include those standard methoad and techniques for each mode of transportation that the unit could be expected to employ. An extract of a battalion SOP that prescribes foot marching procedures is provided herein.

APPENDIX – (FOOT MARCHES) TO ANNEX – (OPERATIONS) TO – BATTALION – INFANTRY TACSOP

1. GENERAL

This appendix is published to standardize foot march procedures and techniques. All information applies to tactical foot marches along improved roads when the enemy situation does not indicate enemy attack. This appendix will be modified only by order or directives from this headquarters. All foot marches conducted by units of this battalion have the implied purpose of arriving at the prescribed destination at the appointed time with soldiers who can immediately accomplish their assigned tactical mission.

2. FACTORS AFFECTING THE MARCH

a. Plans and Orders: The commander involved in a march prepares complete, accurate, and realistic march plans in accordance with guidance contained in paragraph 3 of this appendix. These plans will be translated into timely orders and provide for security, control, minimizing of march losses, and the uninterrupted march of the unit.

b. March Discipline: Specific areas of concern are maintaining formations, distances between men and elements, and speed; and the effective use of cover and concealment, and security. Additionally, commanders will ensure that water, light, noise, and communications disciplines are rigidly enforced to include:

(1) *Water discipline:*

(a) Drink only treated water from approved sources.
(b) Drink water often. Great quantities of water should be drank before, during, and after the march.

(c) Drink water slowly to prevent cramps or nausea.
(2) *Light discipline:* Unshielded light sources are not permitted during the hours of darkness. No smoking at night.
(3) *Noise discipline:* The chain of command inspects every soldier participating in a foot march to eliminate noise by the proper wearing of equipment.
(4) *Communications discipline:* Radio traffic must be reduced to a minimum through the use of arm-and-hand signals and messengers.
(5) *Code words:* Code words are given to indicate five minutes until break, begin break, and three minutes to march.

3. FOOT MARCH PROCEDURES

a. Organization:

(1) *Reconnaissance.* Battalion conducts the following, supplemented as required by personnel from units:
(a) Route reconnaissance.
(b) Reconnaissance of assembly, bivouac, and staging areas.
(c) Traffic control.
(2) *Reconnaissance party.*
(a) Commander-reconnaissance or scout platoon leader.
(b) Composition of reconnaissance party designated separately for each march to include:
1. Reconnaissance element from reconnaissance or scout/platoon.
2. Traffic control element as designated.
3. Engineer element, if available, from attached or supporting engineer unit.
4. NBC reconnaissance and survey team.

(3) *Quartering party.*

(a) Commander–HHC commander.

(b) Composition-communication officer and communication platoon elements; necessary security personnel; and representatives from each company (to include NBC personnel), the S4 section, and the medical platoon.

(4) *Foot marches* (walking elements).

(a) Units are organized into march units and, if necessary for control, into march serials and march columns.

(b) Organization: The battalion is organized with companies as follows —

Rifle company – 4 platoons (includes composite Co HQ Pit).

Headquarters and headquarters company – 1 composite platoon (when directed) march as second march unit; co (-) with motorized elements.

(c) Attachments: Frontline ambulance to each company.

(d) Motorized elements: Integrated with or follow main body by bounds unless shuttle marching is employed.

b. Control Measures: The following control measures will be used by company or larger units and be included in the march order:

(1) Start point (SP) and release point (RP).

(2) Critical points.

(3) Time at which the head of the column is to pass the SP, RP, and critical point.

(4) Date of march.

(5) Order of march.
(6) Route of march.
(7) Distance between men and units.
(8) Column gap.
(9) Pace setter.
(10) Road movement graph (if used)
(11) Road movement table (battalion and brigade only).
(12) Assembly and bivouac areas to include reconnaissance and quartering parties.
(13) Command post (CP location).
(14) Communication to be used during the march.
(15) Lead and trail persons to mark locations with chemical lights during movement.

c. Security: Units provide for continuous all-round security during the march and at halts. Passive and active measures are planned against attack by enemy aircraft or long-range weapons during movement. Reconnaissance elements are employed to determine critical points along the route of march where security should be increased.

d. March Formation: The foot march formation is a staggered column of twos with one file on each side of the road unless otherwise specified.

e. Length of March: The normal length of march for a 24-hour period will be from 20 tp 32 kph marching from 5 to 8 hours at a rate of 4 kph.

f. Rates of March:

	ROADS	CROSS COUNTRY
DAY	4.0 kph	2.4 kph
NIGHT	3.2 kph	1.6 kph

g. Distance between men and units:

		TACTICAL SPEED	
MEN:			
	Day:	5 meters	2 meters
	Night:	3 meters	1 meter
UNITS: PLATOONS			
	Day:	50 meters	25 meters
	Night:	25 meters	15 meters
UNITS: COMPANIES			
	Day:	100 meters	50 meters
	Night:	50 meters	25 meters

h. Night Marches: During night marches control is increased by decreasing the distance between men and units, using connecting files to maintain contact between platoons and companies. Flashlights, lanterns, and pyrotechnics consistent with light discipline can be used for visual communications, depending on the tactical situation. Due to reduced visibility, night marches require the implementation of the following safety measures:

(1) Use off-road trails or routes as much as possible to keep foot troops off roads used by vehicles.
(2) Place marked and equipped guards with reflective or luminous materials at the front and rear of the march units.
(3) Equip moving and static traffic guards with reflective fabric or tape, vests, caps, mittens, or hat

bands. Equip static traffic guards with illuminated or reflective signs, red lights, flashlights, traffic control batons, reflective flags, or lanterns.

(4) Disseminate warning to vehicle operators of the presence of troops on or near the roadway and restrict speeds as appropriate. Instruct vehicle operators on the meaning of luminous warning devices.

i. Forced Marches: For planning purposes, the maximum distances for forced marches are 56 km in 24 hours; 96 km in 48 hours; and 128 km in 72 hours. Full advantage will be taken of the periods when the troops are rested to increase the rate of march, if necessary. Rest periods are taken to ensure the arrival of the unit in effective fighting condition.

j. Halts: A 15-minute halt is made after the first 45 minutes of marching. Thereafter, a 10-minute halt is made every 50 minutes. Observation posts and local security are established during halts. Halfway through the march, a 15-minute rest stop is taken in which all men change their socks. Soldiers move to the sides of the road, seek cover and concealment, face outward, loosen their equipment during halts, and aidmen administer medical treatment as needed. Squad leaders check the feet of men and ensure feet are powdered and socks changed. Meal halts for A-rations last one hour and MREs 30 minutes.

k. Straggler Control: Stragglers who cannot meet or maintain the pace set for the foot march are the responsibility of his immediate chain of command. Fire team leaders, squad leaders, and ultimately the platoon leader must ensure the soldier is linked up with the straggler control party at the rear of the marching column. Information concerning the number of soldiers who fall out, their standard name lines, and medical disposition must be relayed through the chain of command to maintain accountability of personnel an equipment. If individuals who fall out are carrying mission-essential equipment, this equipment must be transferred to other unit members.

APPENDIX B

EXAMPLE OF A BATTALION ROAD MOVEMENT OPERATIONS ORDER

The purpose of an OPORD is to provide clear, concise information and instructions to subordinates so that they may accomplish the mission within the framework of the commander intent. The OPORD must clearly state all required information for the unit to perform the assigned treks. The tasks must be understood for a road movement order since they may be followed by a tactical operation. The five-paragraph OPORD allows for changes to the basic operation to be quickly disseminated. Information for a road movement may be included in the OPORD for a larger operation. therefore, the information is presented as a road movement annex. (Road movement annexes are discussed in detail in FM 55-30.)

(Classification)

(No change from oral orders)

Copy No._of_copies
1st Bn, 66th Inf
MIDLAND (FM980876),
GEORGIA 151000R Apr
19_PRT8

OPERATION ORDER 7
Reference: Map, series v661, GEORGIA, Sheet 4882,
COLUMBUS, Edition 1,1:50,000.

Time Zone Used Throughout the Order: ROMEO
Task Org: Annex A, Road Movement Table.
Annex B, Strip Map (omitted).

1. SITUATION

a. Enemy Forces:

(1) Enemy has withdrawn across MULBERRY Creek (FM9918).

(2) Indications are that enemy will continue to delay on successive positions.

b. Friendly Forces: 1st Bde moves by foot and motor commencing at 160700 Apr to assembly area vicinity MIDLAND (GM0106).

c. Attachment and Detachments: None.

2. MISSION

1-66 Inf moves by foot and motor commencing 160700 Apr to occupy assembly area vicinity FLATROCK (FM9903).

(Classification)

(Classification)

3. EXECUTION

a. Concept of Operation: 1-66 Inf moves by foot in four march units, crossing SP at 160700 Apr and closing new area by 161209 Apr. Motorized elements follow main body by bounds.

(1) Maneuver order of march is MU1, TACCP, MU2, MU3, main CP, MU4, trail party/trains.

(2) Fire support. Priority of 81-mm mortars is to MU1 then to MU2. The 81-mm mortar priority target is allocated to MU1.

(3) Counterair. Priority of protection to main body then to the trail party.

(4) Engineering. Priority of effort is to mobility then survivability. Priority of support is to MU1.

b. March Unit 1.

c. March Unit 2.

d. March Unit 3.

e. March Unit 4. Pick up all road guards and stragglers

f. Coordinating instructions:

(1) Reconnaissance Party assembles Bn Cp 151500 Apr and departs 151600 Apr. Returns NLT 152400 Apr.

(2) Quartering party assembles at the Bn CP at 160300 Apr. Departs 160400 Apr. Quartering Party will post road guards at all checkpoints.

(3) Formation: column of files on each side of road.

(4) Annex A, Road Movement Table.

(5) Annex B, Strip Map (omitted).

4. SERVICE SUPPORT

a. Materiel and Services:

Supply: Cl I. Individuals carry 1 MRE for noon meal 16 Apr.

(Classification)

(Classification)

Transportation: One HMMWV attached to each march unit for straggler control. Four HMMWVs attached to the Scout Platoon for Reconnaissance Party duties. Upon return of Reconnaissance Party, HMMWVs will be attached to quartering party. One 5-ton cargo truck will be attached to March Unit four for road guard police and straggler control.

b. Medical: March collecting points will be located at checkpoints 1 and 6. (Annex A, Road Movement Table).

5. COMMAND AND SIGNAL

a. Command: Command Group and March CP at head of March Unit 2.

b. Signal:

(1) SOI, Index 1-7.

(2) Listening silence except for emergencies.

Acknowledge.

THOMPSON
LTC

Annexes: A - Road Movement Table
B - Strip Map (omitted)

Distribution A

OFFICIAL

SELLERA

S3

(Classification)

(Classification)

(No change from oral orders)

Copy No._of_copies
1st Bn 66th Inf
MIDLAND (FM980876),
GEORGIA 151000S Apr
19_PRT8

Annex A (Road Movement Table) to OPORD 7
Reference: OPORD 7
1. Rate of March: 4 kph.
2. Distances: Between men, 5 meters; between pit, 50 meters; between co, 100 meters.
3. Halts: SOP.
4. Routes: MOYE Road; BUENA VISTA Road; SCHATULGA Road.
5. Route Classification Restrictions: None.
6. Critical Points:
  a. SP: Jet MOYE Rd and 2d ARMD DIV Rd (FM 988882).
  b. RP: Jet SCHATULGA Rd and WARM SPRINGS Rd FM 998035).
  c. Other Critical Points:
    Jet MOYE Rd and ST MARYS Rd (FM 987696) (CP1).
    Jet MOYE Rd and STEAM MILL Rd (FM 998908) (CP2.
    Jet MOYE Rd and BUENA VISTA Rd (FM 998940) (CP3).
    Jet BUENA VISTA Rd and SCHATULGA Rd (GM 001942) (CP4).
    Jet SCHATULGA Rd and FORREST Rd (GM 001963) (CP5).
    Jet SCHATULGA Rd and MACON Rd (GM 000002) (CP6).

(Classification)

(Classification)

7. Main Route to Start Point: MOYE Rd.
8. Main Route from Release Point: Unnamed Trail (as designated by guides).
9. March Units: (See chart.)

(Classification)

(Classification)

MARCH UNIT	ORG & CDR	PT (MIN)	CHECK POINTS	ARRIVAL TIME	CLEARANCE TIME	REMARKS
1	Co A (CPT May)	7	SP	160700 Apr	160707 Apr	Includes 1/4-ton Ambulance
			CP 1	0711	0718	
			CP 2	0733	0740	
			CP 3	0836	0843	
			CP 4	0838	0845	
			CP 5	0910	0917	
			CP 6	1010	1017	
			RP	1050	1057	
2	HHC (-) (CPT Perkins)	3	SP	0724	0727	Includes 1/4-ton Ambulance and Bn March CP and Command Gp
			CP 1	0735	0738	
			CP 2	0757	0800	
			CP 3	0900	0903	
			CP 4	0902	0905	
			CP 5	0934	0937	
			CP 6	1034	1037	
			RP	1114	1117	
3	Co B (CPT Gilbert)	7	SP	0748	0755	Includes 1/4-ton Ambulance
			CP 1	0759	0806	
			CP 2	0821	0828	
			CP 3	0924	0931	
			CP 4	0926	0933	
			CP 5	0958	1005	
			CP 6	1058	1105	
			RP	1138	1145	
4	Co C (CPT Russel)	7	SP	0812	0819	Includes 1/4-ton Ambulance
			CP 1	0823	0830	
			CP 2	0845	0852	
			CP 3	0948	0955	
			CP 4	0950	0957	
			CP 5	1022	1029	
			CP 6	1122	1129	
			RP	1202	1209	

Chart: control of movement.

(Classification)

APPENDIX C

FOOT CARE

Foot hygiene and sanitation are extremely important since feet are enclosed in heavy rigid footwear during most working hours and are constantly inaction. Foot care involves good hygiene measures such as bathing frequently, using foot powder, wearing properly fitted footwear to allow for ventilation, and correctly trimming toenails.

Section I
FOOT HYGIENE

The care of minor foot ailments should be given the utmost attention. Many major conditions requiring hospitalization and disability have resulted from neglected or maltreated minor conditions.

C-1. CONDITIONING

Conditioning is accomplished by progressively increasing the distance to be marched from day to day. Marching is a good way to strengthen the feet and legs; running alone will not suffice. The arch, ankle, and calf can be conditioned by performing simple exercises—for example, rising high on the toes or placing the feet on towels and using the toes to roll the towel back under the arch.

C-2. PREVENTIVE MEASURES

Certain preventive measures can be implemented to avoid painful foot problems.

a. **Before Marches.** Trim toenails at least every two or three weeks, depending upon individual needs. Cut toenails short and square, and straight across (Figure C-1). Keep feet clean and dry, and use foot powder. Wear clean, dry, unmended, good-fitting socks (preferably cushion-soled) with seams and knots outside.

A nylon or polypropylene sock liner can reduce friction and add protection. Carry an extra pair of socks. Carefully fit new boots. When getting used to a new pair of boots, alternate with another pair; tape known hot spots before wearing.

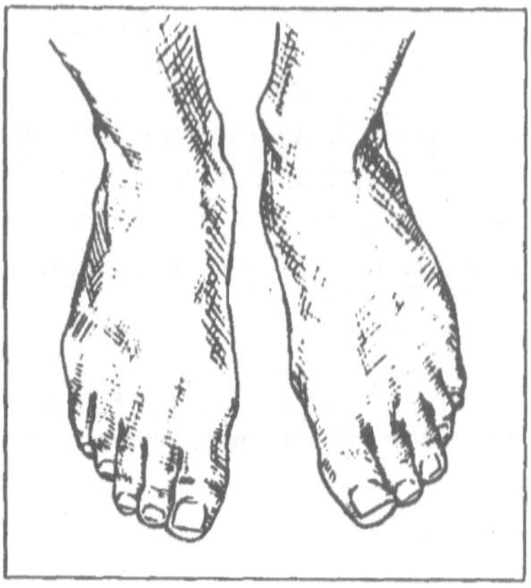

Figure C-1. Trimming of toenails.

b. **During Halts.** Lie down with the feet elevated during each halt. If time permits, massage the feet, apply foot powder, change socks, and medicate blisters. Cover open blisters, cuts, or abrasions with absorbent adhesive bandages. Obtain relief from swelling feet by slightly loosening bootlaces where they cross the arch of the foot.

c. **After Marches.** Repeat procedures for the care of feet, wash and dry socks, and dry boots. Medicate blisters, abrasions, corns, and calluses. Inspect painful feet for sprains and improper fitting of socks and boots. Feet can develop red, swollen, tender skin along the sides of the feet from prolonged marching, which could become blisters. Therefore, feet require aeration, elevation, rest, and wider footwear. Prevent major foot problems by keeping the feet clean. The formation of blisters and abrasions with dirt and perspiration can cause infection and serious injury. If possible, give the feet a daily foot bath. In the field, cool water seems to reduce the sensation of heat and irritation. After washing, dry the feet well.

Section II
FOOT INJURIES

Many foot injuries can occur from foot marches, but only the most common are discussed herein.

C-3. BLISTERS AND ABRASIONS

Common causes of blisters and abrasions are improperly conditioned feet, ill-fitting footwear and socks, improperly maintained footwear, heat, and moisture. They are normally caused by friction or pressure, as opposed to impact.

a. To clean a blister, wash gently around it with soap and water, being careful not to break the skin (Figure C-2). If unbroken, use a sterilized needle or knifepoint to prick the lower edge of the blister to remove fluid. (To sterilize needle or knifepoint, hold in a flame.) Do not remove the skin; cover the blister with an absorbent adhesive bandage or similar dressing, extending beyond the edge of the blister. After applying the dressing, dust the outside of the dressing and entire foot with foot powder.

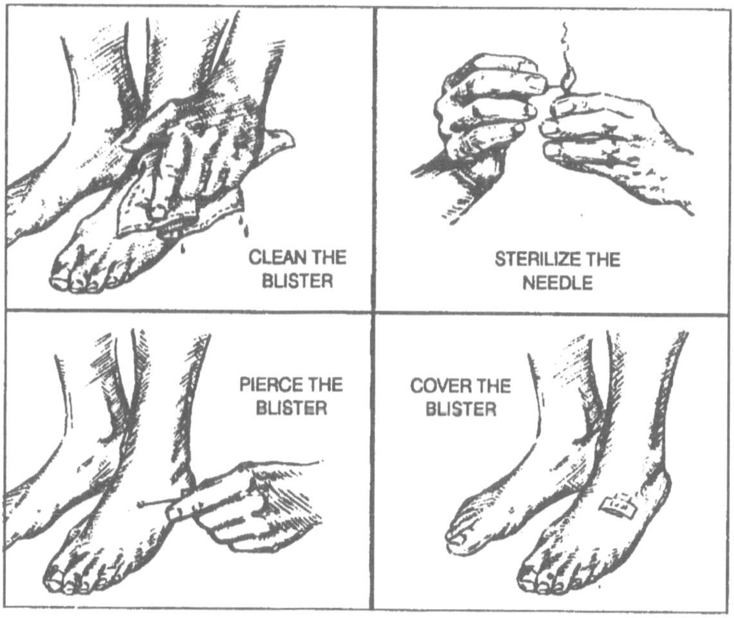

Figure C-2. Treatment of a foot blister.

b. Use just enough foot power since it can harden and become irritating. Foot powder lessens friction on the skin and prevents the raw edges of the adhesive plaster from adhering to socks. The adhesive plaster should be smooth so it can serve as a "second skin." Check the blister periodically for proper drying. After the blister has dried, remove the adhesive plaster. Carefully inspect the foot for other problem areas that are red and tender that may need the protection of an adhesive plaster. Cover abrasions and cuts on the foot with absorbent adhesive bandages for rapid healing. In an emergency, medical personnel can inject tincture of benzoin into a blister to prevent further abrasion and loss of skin.

C-4. PERSPIRATION PROBLEMS

When feet perspire, the secretion decomposes and causes a foul odor. The skin between the toes usually becomes white and soft, rubs off easily, and is prone to abrasions. Treatment consists of washing and thoroughly drying the feet, and carefully painting the affected area with a cotton swab and the following solution

Formalin – one part
Rubbing alcohol – nine parts.

This solution should be allowed to dry. If the skin begins to burn, the excess solution should be washed off. It should be kept out of abrasions and cuts since it can cause severe pain. The entire area of the foot to include the ankle is painted. The areas around the heel and instep, and between toes should be treated since they are main trouble spots. The solution should be applied once daily until the perspiration is halted and the skin becomes hardened.

C-5. ATHLETE'S FOOT

Athlete's foot usually occurs between the toes, on the sole of the foot, and at points of contact between skin and footwear. This and other mild chronic cases of fungus infection may respond to daily foot powder applications. If fungicidal ointment is available, it can be used in addition to foot powder. Ointment should be used as directed and while the feet are at rest. If applications of foot powder and ointment do not heal the infection, an aidman or surgeon should be consulted.

C-6. FROSTBITE

Frostbite is the freezing of a body part due to exposure to below-freezing temperatures, and it is classified as either superficial or deep. It is a constant hazard in operations performed at freezing temperatures, mainly when accompanied with strong winds. Normally, a cold sensation occurs that is followed by numbness and then a tingling, stinging, or aching, or even a cramping pain. The skin first turns red and then yellowish, pale gray, or waxy white.

a. Prevention of frostbite, or stopping it in its early stages, is easier than thawing and caring for frozen flesh. Good-fitting clothing and properly worn equipment avoid interference with blood circulation, which could reduce the amount of heat delivered to the extremities. To prevent severe frostbite –

(1) Enough clothing must be worn for protection against cold and wind. The face must be protected during high winds and during exposure to aircraft propeller blast.

(2) Clothing and the body must be kept dry. To avoid perspiring when performing heavy work in the cold, soldiers should remove outer layers of clothing and then replace them when work is finished. Socks should be changed when the feet become moist.

(3) Cold metal should not be touched with the bare skin in extremely low temperatures. To do so could mean loss of skin.

(4) Adequate clothing and shelter must be provided during inactive times.

(5) The face, fingers, toes, and ears should be exercised or massaged to keep them warm and to detect any numb or hard areas.

(6) The buddy system should always be used. Soldiers should find partners and observe each other for signs of frostbite and for mutual aid if frostbite occurs. Any small frozen spots should be thawed immediately, using bare hands or other sources of body heat.

b. Some cases of frostbite may be superficial, which involves only the skin. If freezing extends below the skin, it demands more involved treatment to avoid or lessen the loss of the body part (fingers, toes, hands, or feet). Often there is no pain, so soldiers must observe each other for signs. Since it is difficult to distinguish between superficial and deep frostbite, a soldier should assume the injury to be deep and therefore serious. If numbness has been for a short time, the frostbite is probably superficial.

c. For treatment of superficial frostbite, the following measures must be adhered to:

(1) Cover cheeks with warm hands until the pain returns.

(2) Place uncovered frostbitten fingers under the opposing armpits next to the skin.

(3) Place bared frostbitten feet under clothing and against the chest of a companion.

(4) Do not rewarm by such measures as massage, exposure to open fires, cold-water soaks, or rubbing with snow.

(5) Be prepared for pain when thawing occurs.

d. For treatment of deep frostbite (freezing injury), the following measures must be adhered to:

(1) If freezing is considered deep, do not attempt to treat the injury in the field. This only causes increased pain and invites infection, greater damage, and even gangrene. Quickly evacuate injured personnel to a hospital or aid station.

(2) Protect the frozen body part from further injury and do *not* try to thaw it by rubbing, bending, or massaging.

(3) Do *not* rub body part with snow or place in cold or warm water; do *not* expose to hot air or open fires; and do *not* use ointments or poultices.

e. Soldiers should not walk on feet after thawing; it is safer to walk on frozen feet. However, thawing may occur spontaneously during transporting to a medical facility. This cannot be avoided since the entire body of the injured soldier must be kept warm.

f. Soldiers who have sustained any form of cold injury are more susceptible to a reoccurrence.

C-7. TRENCHFOOT

Trenchfoot is a thermal injury due to exposure to severe cold-weather conditions in a damp or wet environment in temperatures between 32 and 50 degrees F. Attributing causes include immobility of the limbs as in sitting or standing, insufficient clothing, and constriction of body parts due to boots, socks, and other garments.

a. Trenchfoot is similar to gradual frostbite since the primary causes are the same. The only difference is in the degree of cold. In the early stages of trenchfoot, feet and toes are pale, and they feel cold, numb, and stiff. Walking becomes labored. If preventive action is not taken at this stage, the feet will swell and become painful. In extreme cases of trenchfoot, the flesh dies and amputation of the foot or leg may be needed. Because the early stages of trenchfoot are not painful, soldiers must be alert to prevent it.

b. Socks and boots should be cleaned and dried daily, and feet should be dried soon after being wet. If soldiers must wear wet boots and socks, the feet should be exercised by wiggling the toes and bending the ankles, and they should be warmed with the hands. Then foot powder should be applied and dry socks put on.

c. In treating trenchfoot, the feet should be handled gently; they should not be rubbed or massaged. If needed, feet can be cleaned carefully with plain white soap and water, dried, elevated, and left exposed. While it is best to warm the patient, the feet should always be at room temperature. The patient should be carried and not allowed to walk on damaged feet.

C-8. IMMERSION FOOT

Immersion foot is a form of injury that follows prolonged immersion of the feet in water that is not cold enough to cause freezing or frostbite. It can occur after exposure in subtropical waters. Clinically and pathologically, immersion foot is like trenchfoot since its cause is the same – lowering the temperature of the body part involved. It is associated with dependency (legs and feet down as in sitting or standing) and immobility of the lower extremities, and with constriction of the limbs by clothing or shoes. Other important factors are: body cooling due to wind, total immersion, inadequate protective clothing, illness, and starvation. The treatment for immersion foot is the same as for trenchfoot.

C-9. STRESS FRACTURES AND MUSCLE INJURIES

Once stress fractures have occurred, they must be allowed time to heal. The affected area must rest for two or three weeks until the pain is gone, followed by a slow return to activity to avoid recurring injury. Personnel who have had an injury are more likely to be injured again. The best form of treatment for this type injury is prevention. This can be accomplished through a conditioning program and by ensuring major muscle groups are properly stretched and warmed up before marching.

Section III
CARE OF FOOTWEAR

Boots must be dried after use to avoid loosing shape and hardening of the leather. This can be done by placing a warm cloth in the boot or by any method that avoids rapid drying. To prevent moist leather from freezing during winter, boots should be placed inside a sleeping bag or used as a headrest.

C-10. BOOTS

Two important factors in fitting boots are: the space between the end of the great toe and the toe of the boot should be the width of the thumb; and, in the unlaced boot, there should be enough space under the lower edge of the tongue to insert an index finger (Figure C-3).

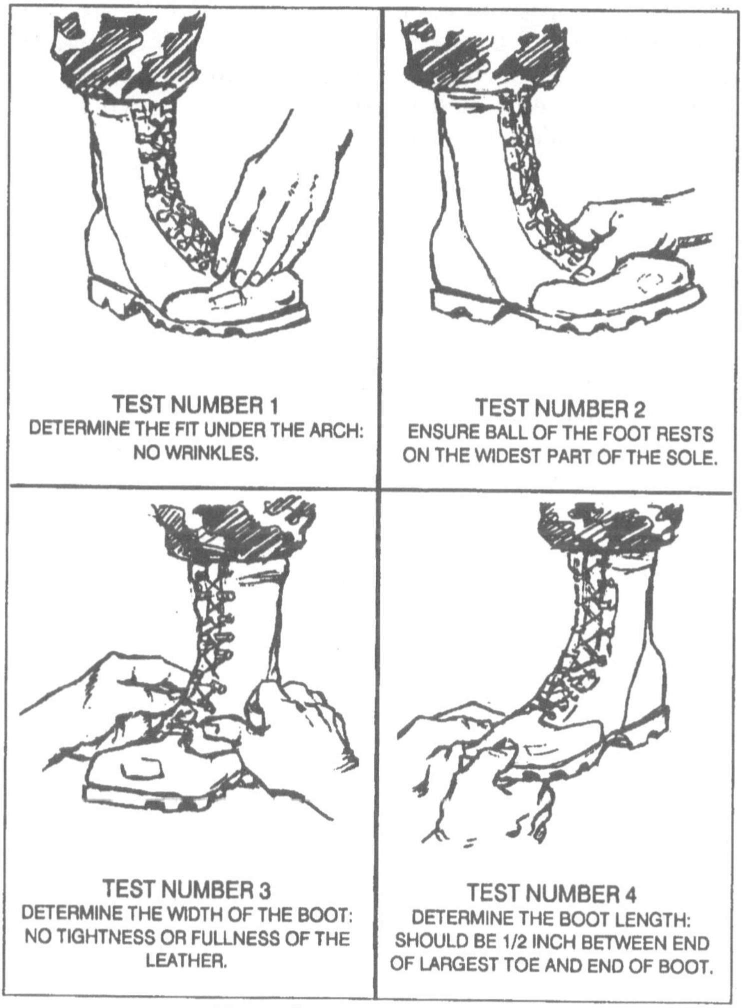

Figure C-3. Fitting of boots.

a. Poorly fitted boots can cause blisters, abrasions, calluses, and corns. Pressure is caused by boots being too small; friction is caused by boots being too large. If the tops of the toes are involved, the cap is too low or too stiff. If the ends of the toes are affected, the boot is too short or too loosely laced. If the sides of the big and little toes become irritated, the boot is too narrow. Irritation at the heel is caused by boots being too long, too loosely laced, or too wide a heel space.

b. Proper lacing of boots not only prevents blisters but also prevents improper blood flow in the foot. Laces can assume a seesaw action, which can produce along blister across the instep. To prevent blistering, lacing over the instep can be avoided. If possible, broad laces should be used and an extra pair should be carried.

C-11. SOCKS

To check the fit of socks, a soldier should stand with his weight evenly distributed on both feet. If the socks fit correctly, no tightness or fullness should exist (Table C-1). The wool cushion-sole sock is best because it offers good foot protection.

a. Soldiers should allow 3/8 of an inch for shrinkage of new socks. Those that are too large wrinkle inside the shoe, rub the feet, and cause blisters and abrasions. Socks that are too small wear quickly and reduce blood flow in the foot. When wearing two pairs of socks, soldiers should wear an outer pair at least a half-size larger than usual. Socks must be changed daily–dirty socks are conductors of heat and allow warmth to escape. They should be washed in lukewarm water to preserve the fiber of the sock since hot water can cause them to shrink.

b. When socks become damp, they can be dried by placing them inside a shirt next to the body. Socks should be completely dry before wearing. If it is not possible to wash the socks, they should be changed; the dirty socks should be dried and kneaded with the hands to remove dirt and hardness.

BOOT SIZE	5-5 1/2	6-6 1/2	7-8	8 1/2-9	9 1/2-10 1/2	11-11 1/2	12-13	13 1/2-14	14 1/2-15
SOCK SIZE*	10	10 1/2	11	11 1/2	12	12 1/2	13	13 1/3	14

*COTTON SOCKS ONE-HALF SIZE SMALLER.

Table C-1. Proper sizes for wool socks.

APPENDIX D

ASSEMBLY AREAS

The battalion occupies an assembly area for security while preparing for future operations. Preparations include reorganizing issuing orders, receiving and issuing supplies, and maintaining vehicles and equipment. Occupation of an assembly area can be directed by a higher commander or by the battalion commander. (See FM 7-20 for a detailed. discussion on the occupation of assembly areas.)

D-1. CHARACTERISTICS

A force in an assembly area is given some security by being behind friendly lines, and by having another friendly element between it and the enemy. Regardless, the battalion must defend itself should the enemy break through, bypass forward defenses, drop airborne or air assault forces, or be seen from the air. It normally uses the same techniques used in the perimeter defense. The assembly area should provide –

- Concealment from air and ground observation.
- Cover from direct fire.
- Space for dispersion against massed chemical or nuclear fires.
- Adequate entrances, exits, and internal routes.
- Good drainage and soil conditions that support battalion or attached vehicles.

D-2. QUARTERING PARTY

Before the main body leaves the assembly area, the march commander sends a quartering party (or advance party) to the forward assembly area. The reconnaissance and the quartering parties do not travel as part of the march column. They precede the main body and move by infiltration.

a. A quartering party normally comprises an OIC (HHC commander, XO of the HHC, or battalion S1), security element, communications and medical personnel, and required staff section and subordinate unit personnel. Its mission is to reconnoiter the new area and to guide march elements to the assembly area. The commander of the quartering party must be told the route, order of march, and estimated time of arrival of the main body. A battalion quartering party consists of the quartering parties from each subordinate company. A company quartering party consists of one element from each platoon.

b. The quartering party should have enough guides, markers, and pioneer tools to improve the new area. As march elements clear the RP, quartering party members waiting in covered and concealed positions move out to guide elements to selected areas without halt.

c. To reduce being seen by the enemy during occupation, the quartering party reconnoiters. Then, it organizes the assembly area before the battalion arrives. The party ensures the area has the required features. It selects areas for each company, CP, and CS and CSS element. The party then guides arriving elements into position, This avoids stopping moving units on an exposed route of march.

D-3. ORGANIZATION

The assembly area is organized by assigning companies sectors of the battalion perimeter or dispersed company assembly areas (Figure D-1). It must allow for good dispersion of all elements of the battalion.

a. Security can be augmented by visual observation, sensors, and surveillance devices. Road security is the duty of the company in whose sector the units pass through. Contact points for units can also be named to help in coordinating security. All exits and entrances to the assembly area are strictly controlled. OPs cover key terrain features and avenues of approach.

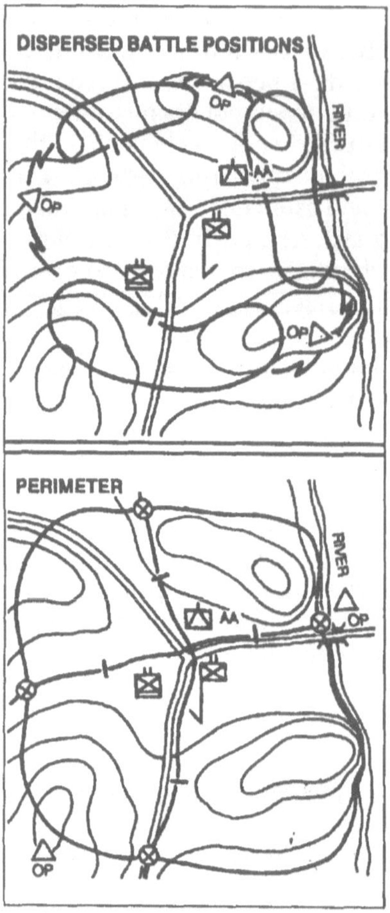

Figure D-1. Assembly areas.

b. The scout platoon may be tasked to reconnoiter routes of movement to counterattack positions, defensive positions, or passage lanes. It may also provide security by setting up OPs, roadblocks, or traffic control points.

c. CS elements are positioned with units they are to support. They are located to provide support to all elements of the battalion.

d. The battalion CP and trains are centrally located for security, which helps to plan, issue orders, distribute supplies, and so on.

e. Communication between elements is by wire or by messenger. Radio is used only in an emergency.

f. Company assembly areas should be large enough to allow for dispersion and for cover and concealment from observation and enemy direct fires. Company positions within the battalion assembly area should aid movement for future operations.

g. Placement of mortars should consider minimum employment distance. This may result in mortars positioned in adjacent company sectors where emplacement must be coordinated. Mortars are normally employed with two-thirds of their range forward. This covers the center of the company sector as the main direction of fire.

h. The quartering party must prepare for occupation of an assembly area during limited visibility. The vital handoff occurs at the RP. Coordination allows for a smooth passage of the march unit through the RP without halts. Guides meet the march unit at the RP. They lead the unit along a marked route to the assembly area. Subunit guides, using planned colored lights or flash recognition signals, link up with platoons/sections and lead them to prepared sectors. Individual, vehicle, crew, or squad positions are marked with stakes, chemical lights, engineer tape, or prelaid communication wire.

GLOSSARY

Section I
ACRONYMS AND ABBREVIATIONS

ADA	air defense artillery
ALICE	all-purpose, lightweight, individual, carrying equipment
APC	armored personnel carrier
AR	Army rgulation
ARTE	Army training and evaluation program
AT	arrival time
BMO	battalion motor office
bn	battalion
CLOHE	combat load handling equipment
cm	centimeter
COMMZ	communications zone
CP	command post
CS	combat support
CSS	combat service support
DA	Department of the Army
EXTAL	extra time allowance
F	Fahrenheit
FM	field manual
FRAGO	fragmentary order

HHC	headquarters and headquarters company
HMMWV	high-mobility, multipurpose, wheeled vehicle
HQ	headquarters
IAW	in accordance with
kg	kilogram
km	kilometer
kmih	kilometers in the hour
kph	kilometers per hour
lb	pound
LBE	load-bearing equipment
LCE	load-carrying equipment
LGTHCOLM	length of column
METT-T	mission, enemy, terrain, troops and time available
mih	miles in the hour
MLC	minimum-load configuration
mm	millimeter
MOPP	mission-oriented protection posture
mph	miles per hour
MRE	meal, ready to eat
NATO	North Atlantic Treaty Organization
NBC	nuclear, biological, chemical
NL	Netherlands
OP	observation point
OPORD	operation order
plt	platoon
POL	petroleum, oils and lubricants
PT	pass time
RP	release point

S1	adjutant
S3	operations and training officer
S4	supply officer
SLOHE	sustainment load handling equipment
SOI	signal operations instructions
SOP	standing operating procedure
SP	start point
STANAG	standardization agreement
TACSOP	tactical standing operating procedures
TC	training circular
TDIS	time distance
TRADOC	Training and Doctrine Command
US	United States
vpk	vehicles per kilometer
vpm	vehicles per mile
XO	executive officer

Section II
DEFINITIONS

Approach March Load. Items of environmental protection, threat protection, and mission load selected according to METT-T for approach marches where contact with the enemy is unlikely. The average weight of approach march loads in a squad should not exceed 72 lb, but individual loads should consider each man's physical capability.

Arrival Time. The time the head of column, or elements thereof, reaches a designated point, line, or object.

Checkpoint. A predetermined point on the ground used as a means of coordinating friendly movement. Checkpoints are not used as reference points in reporting enemy locations.

Clearance Time. The time at which the tail of a column, or elements thereof, passes a designated point, line, or object.

Column. Formation in which elements are placed one directly behind the other.

Combat Load. Mission-essential equipment, as determined by the commander, required for the soldier to fight and survive immediate combat operations (carried by the soldier or on a close support vehicle.) There are two levels of combat load: approach march load and fighting load.

Combat Load Handling Equipment (CLOHE). A resource that stays with the squad/platoon/company on most dynamic operations and carries that part of the combat load not carried by dismounted soldiers.

CLOHE includes any high-mobility vehicle such as a HMMWV, APC, or helicopter. In some circumstances, animals or porters could perform this role.

Column Gap. The space between two consecutive elements proceeding in the same direction on the same route. It can be calculated in units of length or in units of time measured from the rear of one element to the front of the following element.

Completion Time. The time the tail of a column passes the release point.

Contingency Load. All other items of individual and unit equipment not deemed by the commander to be required for ongoing operations. They are stored out of the operational area to be called forward if required under division or corps arrangements.

Critical Point. This is –

a. A key geographical point or position important to the success of an operation.

b. In point of time, a crisis or turning point in an operation.

c. A selected point along a line of march used for reference in giving instructions.

d. A point where there is a change of direction or change in slope in a ridge or stream.

e. Any point along a route of march where interference with a troop movement may occur.

Distance. This is –

a. The space between adjacent individuals measured in any direction.

b. The space between adjacent men, vehicles, or units in a formation measured from front to rear.

Fighting Load. Essential combat items of environmental protection, threat protection, and mission loads required to achieve success once in contact with the enemy, consisting of clothing worn, LBE/LCE, weapon, ammunition, and water. Fighting loads should be kept to an absolute minimum; individual loads of assaulting troops should not exceed 48 lb.

Length of Column. The length of roadway occupied by a column in movement to include the gaps inside the column from the front to the rear inclusive.

Load-Handling Equipment. Any resource that can be used to carry part of the soldier's load and deliver it to him when and where he requires it.

March Collecting Post. Location on the route of march at which casualties who cannot continue to march are given medical treatment and then moved to medical stations in the rear.

March Unit. Unit which moves and halts at the order of a single commander. It normally corresponds to one of the smaller troop units such as a squad, section, platoon, company, or battery.

Pace Setter. An individual, selected by the column commander, who travels in the lead element to regulate the column speed and to establish the pace necessary to meet the required movement order.

Rate of March. The average number of miles or kilometers to be traveled in a given period to include all ordered halts. It is expressed in miles or kilometers traveled each hour.

Release Point (RP). A well-defined point on a route at which the elements composing a column return under the authority of their respective commanders, each one of these elements continuing its movement towards its own appropriate destination.

Road Movement Graph. A time-space diagram used in planning and controlling both foot and motor marches, and in preparing or checking road movement tables.

Road Movement Table. A composite list showing the general organization and time and space schedule for march movement. It is usually published as an annex to an operation order for road movement.

Road Space. The length of roadway allocated to and actually occupied by a column on a route. Road space is expressed in meters (m) or kilometers (km).

Route Reconnaissance. Careful survey of a route for military purposes. The reconnaissance may be accomplished by ground or aerial elements.

Serial. An element or group of elements within a series that are given a numerical or alphabetical designation for convenience in planning, scheduling, and control.

Speed. The actual rate of the forward movement of a vehicle at a given moment as shown on the speedometer (in kph or mph).

Start Point. A well-defined point on a route at which the elements composing a column begin to be under the control of the commander of this movement. It is at this point that the column is formed by the successive passing, at an appointed time, of each of the elements composing the column. In addition to the principal start point of a column, there maybe secondary start points for its different elements.

Strip Map. Sketch of a march route that maybe drawn to scale and should include identifying landmarks such as towns, bridges, outstanding buildings, or crossroads.

Sustainment Load. Equipment determined by the commander to be required for sustained operations, which is stored at battalion level and brought forward to the soldier as required by the commander under S4 arrangements.

Sustainment Load Handling Equipment (SLOHE). SLOHE is a transportation resource assigned to the battalion for carrying sustainment loads.

Time Distance (TDIS). The time it takes the head of a column, or any single element thereof, to move from one point to another at a given rate of march.

Pass Time. The time that elapses between the moment when the leading element of a column passes a given point and the moment when the last element passes the same point.

REFERENCES

Required Publications

Required publications are sources that users must read in order to understand or to comply with this publication.

Field Manuals (FMs)

21-10	Field Hygiene and Sanitation
21-11	First Aid for Soldier
21-15	Care and Use of Individual Clothing and Equipment
21-26	Map Reading and Land Navigation
21-20	Physical Fitness Training
21-60	Visual Signals
21-75	Combat Skills of the Soldier
101-5-1	Operational Terms and Symbols

Related Publications

Related publications are sources of additional information. They are not required in order to understand this publication.

Army Regulations (ARs)

40-5	Preventive Medicine
40-16	Special Notification-Injury Cases
40-350	Medical Regulating To and Within the Continental United States
310-25	Dictionary of United States Army Terms

Department of the Army Pamphlets (DA Pares)

310-35	Index of International Standardization Agreements

Field Manuals (FMs)

3-3	NBC Contamination Avoidance
3-4	NBC Protection
3-5	NBC Decontamination
5-20	Camouflage
7-10(HTF)	The Infantry Rifle Company (Infantry, Airborne, Air Assault, Ranger) (How to Fight)
7-20	The Infantry Battalion (Infantry, Airborne and Air Assault)
7-30	Infantry, Airborne and Air Assault Brigade Operations
8-35	Evacuation of the Sick and Wounded
19-2	Military Police Traffic Operations
20-32	Mine/Countermine Operations
25-100	Training the Force
31-70	Basic Cold Weather Manual
31-71	Northern Operations
34-1	Intelligence and Electronic Warfare Operations
34-3	Intelligence Analysis
55-30	Army Mortar Transport Units and Operation
90-3(HTF)	Desert Operations (How to Fight)
90-5(HTF)	Jungle Operations (How to Fight)

90-6	Mountain Operations
100-1	The Army
100-5	Operations
100-20	Low Intensity Conflict

Training Circulars (TCs)

21-76	Personal Health Training for Personnel in Basic Training
31-25	Special Forces Waterborne Operations

INDEX

Made in United States
Troutdale, OR
11/03/2026

56847409R00076